Water-Associated Infectious Diseases

Shailendra K. Saxena
Editor

Water-Associated Infectious Diseases

Springer

Editor
Shailendra K. Saxena
Centre for Advanced Research
King George's Medical University
Lucknow
India

ISBN 978-981-13-9196-5 ISBN 978-981-13-9197-2 (eBook)
https://doi.org/10.1007/978-981-13-9197-2

This Springer imprint is published by the registered company Springer Nature Singapore Pte Ltd.
The registered company address is: 152 Beach Road, #21-01/04 Gateway East, Singapore 189721, Singapore

Dedicated
to my ***Parents*** *and* ***Family***
who believed in academics
as the way forward for an intelligent mind
and
to my ***Teachers***
who introduced me the subject
and nurtured my interest in it.

Foreword

Water is vital for the survival of all living things, including people. However, industrial waste, poor hygiene, and ineffective sanitation expose a large part of humanity to the water-related risks of contaminants, toxins, and infectious agents. The impact of water-associated infectious diseases, which are the subject of this book, is strongly dependent on the nature of the pathogens involved and also on broader ecological and climatic factors. Suitable sanitation methods for managing excreta form the crux of prevention, but contaminated water remains the major cause of global public health concern. Certainly, the fact that more than three million people die every year from these infections is shocking and requires sustained attention.

This book introduces the main agents responsible, reviews their epidemiology and evolution, considers the impact of global warming, and sets out the mitigating strategies employed at national and international levels. In addition, it deals with the molecular approaches used to detect the pathogens responsible, various ways in which water can be treated to prevent disease emergence, and the means of managing outbreaks when they occur. The chapters provide insights into the strategies of leading international organizations such as the United Nations (UN), the World Health Organization (WHO), the Centers for Disease Control and Prevention (CDC), and the International Water Association (IWA), as well as highlighting vital national initiatives such as the Swachh Bharat Mission (Clean India Mission), which operates throughout the country under the auspices of the Department of Drinking Water and Sanitation of the Ministry of Jal Shakti, Government of India.

The majority of people affected cannot afford conventional medicines and preventive measures such as vaccines. As a result, this book focuses also on the implementation of sustainable hygienic conditions and discusses robust and reliable policies for providing under privileged people with access to basic sanitation, hygiene, and water. It will be of interest to clinicians, microbiologists, climatologists, pharmaceutical industry representatives and policy-makers, and a timely contribution towards the global elimination of water-associated infectious diseases.

University of Glasgow, Glasgow, UK

Andrew J. Davison
President of the International Committee
on Taxonomy of Viruses

Preface

The escalating number of cases and the emergence of antiquated pathogens have created an immense worldwide burden on humanity. Emerging pathogens can be distributed according to the microbial classification, where viruses and bacteria represent 44% and 30% of overall etiological agents, respectively. The rationale of the emergence of infectious disease can be explained by several factors such as new environment, newer technologies, scientific advancement, changes in human behavior, and vulnerability. Pathogens-associated water-borne diseases are the major cause of mortality and morbidity in the developing and underdeveloped countries. The growing population density is the culprit behind the rising number of outbreaks of water-associated infectious diseases. The changing source of water bodies changes the pathogens related to water-associated infectious diseases.

Newer techniques are required which combines the strategy of accuracy assigned with traditional microbiology and sensitivity related with molecular biology. The plentiful number of conventional- and CAM-based medicines for the treatment of water-associated infectious diseases will supersede the gap of limited number of conventional drugs. This book focuses on the various aspects of implementation of sustainable hygienic conditions and discusses the robust and reliable policies and strategies on a global aspect to provide unprivileged people access with the basic sanitation, hygiene, and water.

All these aspects of the book are imperative for safeguarding human race from more loss of resources and economies due to water-associated infectious diseases. To overcome these issues and fill the gap, we propose to introduce *Water-Associated Infectious Diseases* to provide a readily available resource in this area.

The editor of this book hopes that this work might increase the interest in this field of research and that the readers will find it useful for their investigations, management, and clinical usage. The editor and contributors report no conflict of interest.

Lucknow, India

Shailendra K. Saxena

About This Book

Although we are in the era of the twenty-first century having most of the advanced technologies in hand, water-associated infectious diseases are the reason for the worldwide morbidity and mortality. This book provides a comprehensive overview of water-associated infectious diseases and their linked pathogens with plausible strategies and schemes for their mitigation at national and international level. This book focuses on various aspects of implementation of sustainable hygienic conditions and discusses the robust and reliable policies and strategies on a global aspect to provide unprivileged people access with the basic sanitation, hygiene, and water. All these aspects are imperative for safeguarding human race from more loss of resources and economies due to water-associated infectious diseases. To overcome these issues and fill the gap, we hope *Water-Associated Infectious Diseases* shall provide a readily available resource in this area.

Acknowledgements

This book was conceptualized to focus on water-associated infectious diseases and its management including various aspects of implementation of sustainable hygienic conditions and discusses the robust and reliable policies and strategies on a global aspect to provide unprivileged people access with the basic sanitation, hygiene, and water.

All these aspects are imperative for safeguarding human race from more loss of resources and economies due to water-associated infectious diseases. To overcome these issues and fill the gap, we hope *Water-Associated Infectious Diseases* shall provide a readily available resource in this area. The aim of this book is to acknowledge the potential of water-associated infectious diseases and their remedies.

I am overwhelmed in all humbleness and gratefulness to acknowledge from the bottom of my heart to all the contributors who trusted me and supported this work. I hope they are as proud of this book as I am. I would also like to thank Springer Nature Publisher to consider this book for publication. All the reports cited in this book are taken with proper citation. However, any missed information is just unintentional and explicable.

My research fellows and students are central to all my research and academic work. They are motivating force behind anything constructive we do. They are truly brilliant and have a bright future. I would like to express my special thanks of gratitude to my mentors, teachers, and students who gave me the strength to accomplish this. Also, I would like to thank the colleagues, family, and friends who gave a lot of encouragement and support during the work on this book.

A happy environment at home is essential for any kind of growth, and I thank my family, especially my talented wife and children for the same.

Lucknow, India — Shailendra K. Saxena

Contents

About the Editor

Shailendra K. Saxena is Vice Dean and Professor & Head CFAR at King George's Medical University, Lucknow. His primary research interest is to understand molecular mechanisms of host defense during human viral infections and to develop new predictive, preventive, and therapeutic strategies for them. His work has been published in reputed international journals with high-impact factor. His work has been highly cited by numerous investigators globally and honored by several prestigious national and international awards, fellowships, and scholarships in India and abroad, including various Young Scientist Awards and BBSRC India Partnering Award. In addition, he was named as the Global Leader in Science by *The Scientist* magazine (USA) and International Opinion Leader/Expert involved in the vaccination for JE by IPIC (UK). He has been elected Fellow of The Royal Society of Biology (FRSB) and the Royal Society of Chemistry (FRSC) both from the United Kingdom; The Academy of Environmental Biology, India (FAEB); Indian Virological Society (FIVS); The Biotech Research Society, India (FBRS); and the (European) Academy of Translational Medicine Professionals (FacadTM), Austria. Furthermore, he has been awarded Dr. JC Bose National Award of the Department of Biotechnology (DBT, Ministry of Science and Technology, Government of India) in Biotechnology and has active collaboration with US universities.

Introduction to Water-Associated Infectious Diseases

1

Shailendra K. Saxena, Swatantra Kumar, Amrita Haikerwal, and Vimal K. Maurya

Abstract

Considerable advancement in the area of biomedical research and biotechnology has attributed a new shape to the lifestyle and healthcare. However, emerging and re-emerging infectious diseases have always been a major threat coupled with immense challenges, suggesting the necessity of implication of innovative approaches and advanced technologies. Contaminated water is the predominant cause of human exposure to the infectious etiological agents including pathogens, toxins, and organic and inorganic contaminants. Water safety plans (WSP) have been included in the WHO guidelines for drinking-water quality with several imperative primary objectives.

Keywords

Water · Emerging pathogens · Contaminated water · Water safety plans

1.1 Introduction

The escalating number of cases and emergence of antiquated pathogens have created an immense worldwide burden on humanity. Considerable advancement in the area of biomedical research and biotechnology has attributed a new shape to the lifestyle and healthcare. However, emerging and re-emerging infectious diseases have always been a major threat coupled with immense challenges, suggesting the necessity of implication of innovative approaches and advanced technologies. Emerging pathogens can be distributed according to the microbial classification

S. K. Saxena (✉) · S. Kumar · A. Haikerwal · V. K. Maurya
Centre for Advanced Research, Faculty of Medicine, King George's Medical University, Lucknow, India
e-mail: shailen@kgmcindia.edu

S. K. Saxena (ed.), *Water-Associated Infectious Diseases*,
https://doi.org/10.1007/978-981-13-9197-2_1

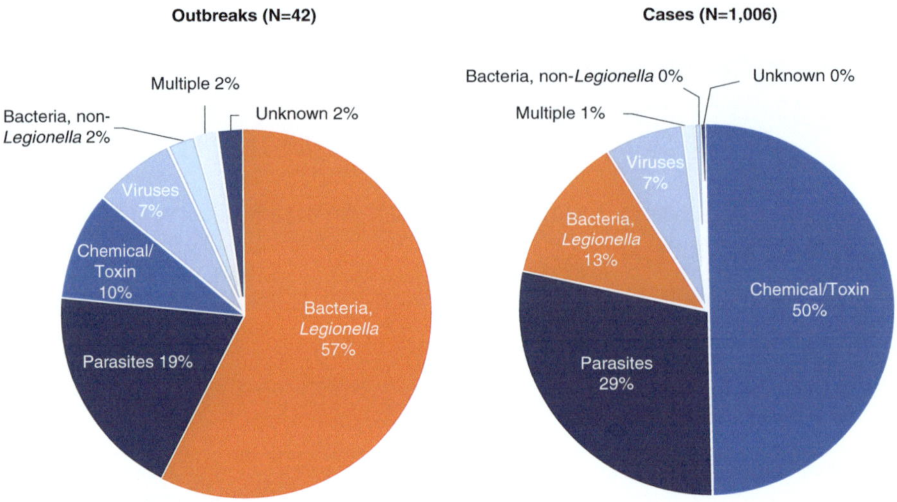

Fig. 1.1 Etiology of drinking water outbreaks and outbreak-related cases (2013–14) [Credit-CDC]

where viruses, parasites and bacteria represent 7%, 19% and 57% of overall etiological agents, respectively (Fig. 1.1) [1]. Socioeconomic development and recent explosion in population growth is exerting a relentless pressure on quantity and quality of water affecting the most imperative life form, mankind [2]. With the recent advancement of multidisciplinary research, the understanding of the correlation between human health and water results in the characterization of previously unidentified pathogens as well as the significance of water-associated pathogens [3]. The rationale of the emergence of infectious disease can be explained by several factors such as new environment, changes in human behavior and vulnerability, emergence of novel technologies, and scientific innovations. Pathogen-associated water-borne infections are the major cause of mortality and morbidity in developed and developing countries [4]. Water is an essential constituent of life, of which oceans comprise 97% and only 3% is available for the living creatures in the form of freshwater. Contaminated water, poor sanitation, hygiene, recreation, industrial waste, and use of unprotected drinking water may describe the higher vulnerability of the human population for water-associated risks [5]. Contaminated water is the predominant cause of human exposure to the infectious etiological agents including pathogens, toxins, and organic and inorganic contaminants [6]. Water safety management depends primarily on hazard identification and ascertaining the availability of preventive measures [7]. Exploitation of suitable sanitation for the unambiguous management of excreta represents the primary barrier to the prevention of outbreak of water-associated diseases [8]. However, water resources contaminated with fecal matter remain the major cause of global public health concern [9]. Numerous techniques are employed for the identification and evaluation of fecal contaminated drinking water which includes simple, economical, and point-of-care tests [10]. The hydrogen sulfide (H_2S) detection method intends to identify the H_2S-producing bacteria commonly associated with fecal contamination [11]. Water safety plans (WSP) have been included in the *WHO guidelines for drinking-water quality* with several imperative primary objectives [12]. The WSP intent to supply good quality of drinking water, eliminate contaminants through various treatment processes, and take preventive steps for reduced level of contamination while in storage [13].

References

1. Jones KE, Patel NG, Levy MA, Storeygard A, Balk D, Gittleman JL, Daszak P (2008) Global trends in emerging infectious diseases. Nature 451(7181):990–993
2. Zimmerman JB, Mihelcic JR, Smith J (2008) Global stressors on water quality and quantity. Environ Sci Technol 42(12):4247–4254
3. Blackburn BG, Craun GF, Yoder JS, Hill V, Calderon RL, Chen N, Lee SH, Levy DA, Beach MJ (2004) Surveillance for waterborne-disease outbreaks associated with drinking water—United States, 2001–2002. MMWR Surveill Summ 53(8):23–45
4. Prüss-Ustün A, Bartram J, Clasen T, Colford JM Jr, Cumming O, Curtis V et al (2014) Burden of disease from inadequate water, sanitation and hygiene in low- and middle-income settings: a retrospective analysis of data from 145 countries. Tropical Med Int Health 19(8):894–905
5. Clasen T, Pruss-Ustun A, Mathers CD, Cumming O, Cairncross S, Colford JM Jr (2014) Estimating the impact of unsafe water, sanitation and hygiene on the global burden of disease: evolving and alternative methods. Tropical Med Int Health 19(8):884–893
6. Richardson SD, Ternes TA (2014) Water analysis: emerging contaminants and current issues. Anal Chem 86(6):2813–2848
7. Hamilton PD, Gale P, Pollard SJ (2006) A commentary on recent water safety initiatives in the context of water utility risk management. Environ Int 32(8):958–966
8. Curtis V, Cairncross S, Yonli R (2000) Domestic hygiene and diarrhoea—pinpointing the problem. Tropical Med Int Health 5(1):22–32
9. Harwood VJ, Staley C, Badgley BD, Borges K, Korajkic A (2014) Microbial source tracking markers for detection of fecal contamination in environmental waters: relationships between pathogens and human health outcomes. FEMS Microbiol Rev 38(1):1–40
10. Ashbolt NJ (2015) Microbial contamination of drinking water and human health from community water systems. Curr Environ Health Rep 2(1):95–106
11. Pathak SP, Gopal K (2005) Efficiency of modified H2S test for detection of fecal contamination in water. Environ Monit Assess 108(1–3):59–65
12. World Health Organization (2011) Guidelines for drinking-water quality, 4th edn. WHO, Geneva. ISBN: 978 92 4 154815 1. http://www.who.int/water_sanitation_health/publications/dwq-guidelines-4/en/. Accessed on 12-3-2018
13. World Health Organization. WSPortal: health through water. water safety plans. Geneva: WHO. http://www.who.int/wsportal/wsp/en/. Accessed on 12-3- 2018

Etiological Agents of Water-Associated Infectious Diseases

2

Swatantra Kumar*, Amrita Haikerwal,
and Shailendra K. Saxena*

Abstract

Based on the origin etiological agents of water-associated infectious diseases are mainly classified into five categories as water-based, water-borne, water-related, water-washed, and water-dispersed. Huge number of diseases are associated with water-borne infections including typhoid fever, cholera, and bacillary dysentery. The principal pathways involved in the transmission of water-borne diseases can be described as 5F concept which are fluids, finger, food, fields, and flies. Water-based etiological agents are mainly worms which spend a part of their life cycle in water bodies. Schistosomiasis is the best known classical example of water-based infectious disease. Water-related infectious diseases are zoonotic in nature where breeding of vectors such as mosquitoes and insects may serve as an initial replication of pathogens and eventually results in the disease transmission from vector to dead-end host. Poor personal and domestic hygiene can be associated with water-washed diseases which are related to clean water deprivation resulting in contact of eye or skin with contaminated water.

Keywords

Water-based · Water-related · Water-washed and Water-dispersed · Vector borne
Fecal contamination · Transmission

---

*Contributed equally as first author

S. Kumar · A. Haikerwal · S. K. Saxena (✉)
Centre for Advanced Research, Faculty of Medicine, King George's Medical University, Lucknow, India
e-mail: shailen@kgmcindia.edu

S. K. Saxena (ed.), *Water-Associated Infectious Diseases*,
https://doi.org/10.1007/978-981-13-9197-2_2

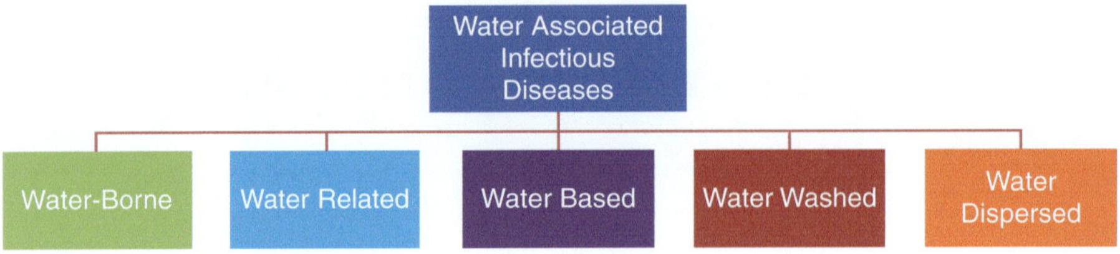

Fig. 2.1 Classification of water-associated infectious diseases

2.1 Introduction

Water-associated infections are the major threats to the population living in the developing countries. Mode of transmission of water-associated infectious diseases can be classified into four categories. Depending upon the process and nature of transmission all the four categories are described as direct transmission, vector-borne, zoonotic, and environmentally mediated [1]. Based on the origin etiological agents of water-associated infectious diseases can be broadly classified into five categories (Fig. 2.1), namely water-borne, water-based, water-related, water-washed, and water-dispersed [2].

2.2 Water-borne

Water-borne infections can be described as pathogens present in contaminated water that can be further classified into four categories: bacteria, protozoa, helminths, and enteroviruses [3]. A huge number of diseases are associated with water-borne infections where primary diseases (Fig. 2.2) are typhoid fever, ascariosis, bacillary dysentery, gastroenteritis, leptospirosis, poliomyelitis, meningitis, hepatitis, amoebic meningoencephalitis, and cholera. Prominent fecal contamination can be detected as the continuous presence of the ample number of microorganisms in the feces. Nearly most of the bacteria present in the human feces are characterized as obligate anaerobes [4] and are present in a higher concentration ($>10^{10}$/g) which is commonly associated with difficulties in their isolation. The most common bacterium present in fecal contaminated water is termed as "coliform" which resembles *E. coli* [5]. Total coliform can be described as gram-negative, facultative anaerobes and spore-producing, rod-shaped bacteria and can ferment lactose into acid and gas within 48 h at 35 °C [6]. The principal pathways involved in the transmission of water-borne diseases can be described as 5F concept which are fluids, finger, food, fields, and flies [7]. Sanitization of the fecal contaminated water resources is poorly investigated with insignificant outcomes but has provided information about the source of disease transmission which can be targeted in effective interventions [8].

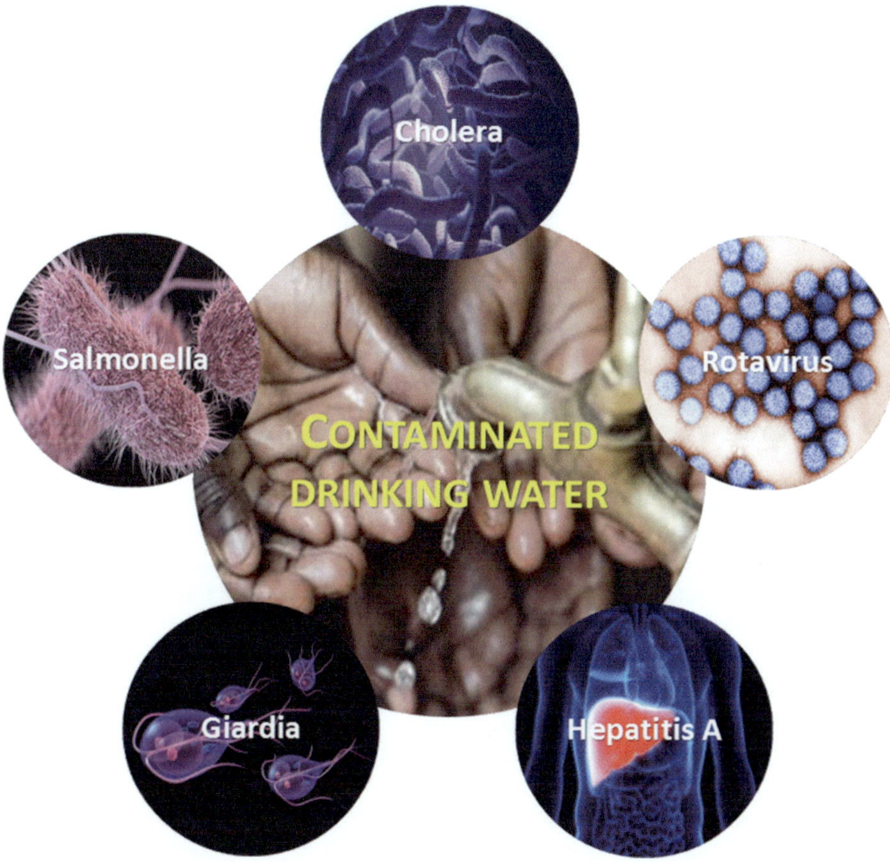

Fig. 2.2 Diseases associated with contaminated water

2.3 Water-Based

Water-based etiological agents are mainly worms which spend a part of their life cycle in water bodies. Schistosomiasis is the best classical example of water-based infectious disease in humans [9]; after internalizing into the human host the larvae reproduce sexually and this results in generation of egg which is passed from urine and stool. Upon reaching the freshwater resources the egg develops into a different larval stage known as *miracidia*, primarily infecting snails where asexual multiplication results in shedding of *Cercariae* and thus completing transmission cycle [10].

2.4 Water-Related

Water-related infectious diseases are zoonotic in nature where breeding of vectors such as mosquitoes and insects may serve as an initial replication of pathogens and eventually results in the disease transmission from vector to dead-end host [11]. The predominant vectors are mosquitoes which can transmit the etiological causative agents to the humans upon biting. Principal vector-borne viral infections include viruses from family *Flaviviridae* [12] comprising Zika virus, West Nile virus, Dengue virus, Japanese encephalitis virus, and Yellow fever virus. Similarly, the predominant parasitic infections based on water-related diseases are malaria, trypanosomiasis, and filariasis.

2.5 Water-Washed

Poor personal and domestic hygiene is mainly associated with water-washed infections which are caused by clean water deprivation and mainly contact with contaminated water. A wide range of diseases such as leprosy, shigella, skin infections, trachoma, ulcers, scabies, and conjunctivitis can be assigned in this category [13].

2.6 Water-Dispersed

Water-dispersed infectious etiological agents are mainly propagated in natural water sources and may enter through respiratory tract into the human body. Pontiac fever (Legionellosis) is the well-known water-dispersed disease caused by *Legionella*. The transmission occurs when a person takes breath in a mist, contaminated with *Legionella* [14]. The most commonly documented source of infection is water used for personal care such as showering, hot tubs, and air-conditioning systems [15].

References

1. Eisenberg JN, Desai MA, Levy K, Bates SJ, Liang S, Naumoff K, Scott JC (2007) Environmental determinants of infectious disease: a framework for tracking causal links and guiding public health research. Environ Health Perspect 115(8):1216–1223
2. White GF, Bradley DJ, White AU (2002) Drawers of water: domestic water use in Africa: public health classics. Bull World Health Organ 80(1):63–73; discussion 61–62
3. Ashbolt NJ (2004) Microbial contamination of drinking water and disease outcomes in developing regions. Toxicology 198(1–3):229–238
4. Sinton LW, Finlay RK, Hannah DJ (1998) Distinguishing human from animal faecal contamination in water: a review. NZ J Marine Freshw Res 32(2):323–348
5. Lamka KG, LeChevallier MW, Seidler RJ (1980) Bacterial contamination of drinking water supplies in a modern rural neighborhood. Appl Environ Microbiol 39(4):734–738
6. Clark JA, Pagel JE (1977) Pollution indicator bacteria associated with municipal raw and drinking water supplies. Can J Microbiol 23(4):465–770

7. Sclar GD, Penakalapati G, Amato HK, Garn JV, Alexander K, Freeman MC, Boisson S, Medlicott KO, Clasen T (2016) Assessing the impact of sanitation on indicators of fecal exposure along principal transmission pathways: a systematic review. Int J Hyg Environ Health 219(8):709–723
8. Boisson S, Engels D, Gordon BA, Medlicott KO, Neira MP, Montresor A, Solomon AW, Velleman Y (2016) Water, sanitation and hygiene for accelerating and sustaining progress on neglected tropical diseases: a new Global Strategy 2015–2020. Int Health 8(Suppl 1):19–21
9. Colley DG, Bustinduy AL, Secor WE, King CH (2014) Human schistosomiasis. Lancet 383(9936):2253–2264
10. Bertuzzo E, Mari L (2017) Hydrology, water resources and the epidemiology of water-related diseases. Adv Water Resour 108:329–331
11. Weaver SC, Barrett AD (2004) Transmission cycles, host range, evolution and emergence of arboviral disease. Nat Rev Microbiol 2(10):789–801
12. Kumar S, Chitti SV, Kant R, Saxena SK (2016) Insights into the immunopathogenesis during Japanese encephalitis virus infection. J Immune Serum Biol 3(1):1–4
13. World Health Organization (1988) Environmental management for vector control: water related diseases and their public health importance. WHO, Geneva. www.who.int/water_sanitation_health/publications/vectcont/en/. Accessed on 12-3-2018
14. Cunha BA, Burillo A, Bouza E (2016) Legionnaires' disease. Lancet 387(10016):376–385
15. van Heijnsbergen E, Schalk JA, Euser SM, Brandsema PS, den Boer JW, de Roda Husman AM (2015) Confirmed and potential sources of Legionella reviewed. Environ Sci Technol 49(8):4797–4815

Evolution and Interplay of Water-Associated Human Pathogens

3

Swatantra Kumar*, Vimal K. Maurya, and Shailendra K. Saxena*

Abstract

Cholera serogroups have been identified which later includes O139 serogroup. Genetic assortment of O139 strain changes the epidemiological status of the cholera and developed strategies for the persistence in competition O1 serogroups. Similarly, heterogeneity responsible for virulence of *Salmonella* has been conventionally attributed to diverse distribution of genetic elements, namely bacteriophages, chromosomal pathogenicity island, transposons, plasmids, etc. Advancement of DNA sequencing and phylogenetic analysis has led to the understanding of the clear evolutionary relationship of various *Shigella* spp. along with *E. coli*. The spatial heterogeneity and intrinsic spatial structure of CF lung selection seem to play a vital role in the diversification of *P. aeruginosa*. Like other commensal bacteria, *Helicobacter pylori* have evolved several mechanisms to evade immune responses. Wide genome scanning, discovery of the low variation regions, and signature of selective sweeps allowed us to identify the various genes of malaria that underwent mutation throughout the course of evolution. The epidemiological distribution dengue serotypes is similar; however, genetically the serotypes are diverse in nature.

Keywords

Serotypes · Genetic assortment · Evolution · Epidemiology · Mutation · Antibiotic resistance

*Contributed equally as first author

S. Kumar · V. K. Maurya · S. K. Saxena (✉)
Centre for Advanced Research, Faculty of Medicine, King George's Medical University, Lucknow, India
e-mail: shailen@kgmcindia.edu

S. K. Saxena (ed.), *Water-Associated Infectious Diseases*,
https://doi.org/10.1007/978-981-13-9197-2_3

3.1 Introduction

In 1854, the first outbreak of water-borne disease, i.e., cholera, was scientifically reported in London. This has led to the understanding rationale of infection which was due to consumption of sewage contaminated water from domestic well [1]. A decade before the germ theory, these findings demonstrated that the contaminated water can harbor microorganism which can affect human health. Since then, the evolution, epidemiology, and transmission of water-borne infectious diseases became the major area of research worldwide. Around 15% of the world's population lives in the water stress area. Scarcity of clean water has led the vulnerable population to struggle for clean water for drinking and other activities. Approximately 1.5 million children die from diarrheal disease annually and 2.5 billion people have no access to adequate sanitation [2]. Contaminated water is the major cause of water-associated infectious diseases which leads to chronic disability, suffering, and death. Ironically, in developing countries, abundant enduring surface water resources as dams, lakes, rivers, irrigation, and streams worsen the possibility of water-borne infections. Evolutionary study of water-associated pathogens requires integration of multidiscipline such as molecular biology, microbiology, and epidemiology [3]. Accurate models are difficult to understand the evolution of water-associated pathogens due to complexity of life cycle and difficulty of characterization. However, understanding the evolutionary trend of water-borne pathogens may lead us to understand the genetic reassortment (genetic exchange between chromosomes), mutation, gene frequencies resulting from natural selection pressure, genetic drift, and development of antibiotic resistance. Major challenges to study the evolution of any pathogens have been categorized as follows: fitness of pathogen, pathogen diversity, impact of coinfection, host-immune diversity, and genetic basis of evolution [4].

3.2 *Vibrio cholerae*

Cholera remains the major threat to the mankind in the form of water-associated infectious diseases and accountable for the millions of deaths in the past 200 years. Detection of the strain can be accomplished by various approaches including phenotypic characteristic and molecular DNA-based technologies specifically entire genomic approach. These methods have been implemented for studying the epidemiology and evolutionary trend of *V. cholerae* [5]. Because of the low discriminatory power of the phenotype fingerprinting such as biotype, serotype, antibiogram and others are conventional typing methods; other methods has developed. The most widely used typing method has been developed which uses the anti-sera of attenuated (heat killed) microbes. A total of 138 *V. cholerae* serogroups have been identified which later includes O139 serogroup. This classification was based on the variation in the sugar components at the surface of the heat-stable somatic "O" antigen bacteria [6]. Later, the *V. cholerae* O1 has been categorized into three serotypes namely Hikojima, Ogawa, and Inaba [7]. Based on the hemolytic properties two of the major biotypes of *V. cholerae* have been characterized, viz. classical and E1 Tor [8]. The low consistency in the hemolytic properties was found to be an incorrect way of categorizing the *V. cholerae*. Later phage typing came into the picture which was adopted to differentiate the classical and E1 Tor where phage susceptibility was used to differentiate the serotypes [9]. Due to the limitation of the restricted number of *Vibrio* phages, consensus typing was not uniform.

Antimicrobial vulnerability has been used to describe the *Vibrio* spp. where susceptibility of the bacteria was measured against a range of antibiotics at various concentrations. Antibiotic resistance profile was generated which used to generate the clusters of the similar strains. Changing the antibiotic resistance profile is the hallmark of the epidemiology of cholera and is associated with the genetic transfer of resistance gene in *V. cholerae* [10]. Although the phenotypic characterization is used to identify the strain nevertheless after the seventh pandemic of cholera these methods have arisen with several limitation of unstable phenotyping, lower sensitivity, and limited specificity. During 1992–1993, the emergence of *V. cholerae* O139 Bengal was related with the massive outbreaks of cholera in Bangladesh and India with a disarticulation of the existing *V. cholerae* O1 strains. Genetic assortment of O139 strain changes the epidemiological status of the cholera and developed strategies for the persistence in competition O1 serogroups. The emergence of *V. cholerae* O139 has provided an evidence of genetic changes which may be associated with the displacement of other serotypes [11]. Environmental and host factors acted as the natural selection process which may define the genetic changes influencing the epidemiology, evolution, and pathogenicity of *V. cholerae* not limited to the endemic regions.

3.3 *Salmonella*

Salmonella enterica is categorized into six subspecies, i.e., I, II, IIIa, IIIb, IV, and VI and further subcategorized based on the expression of distinctive surface antigens (capsular Vi, flagellar H and somatic O) into approximately 2500 serovars [12]. Majority of human-specific virulent strains belong to subspecies I. Heterogeneity responsible for virulence has been conventionally attributed to diverse distribution of genetic elements, namely bacteriophages, plasmids, chromosomal pathogenicity island, transposons, etc. [13]. Recently, loss of gene via gene deletion, truncation, or insertional inactivation has been the crucial factor for evolution of highly pathogenic *Salmonella* species. The course of adaptation of *Salmonella* and its host is conventionally considered to acquire the newer genetic elements which codes for several pathogenic factors [14]. One type of mutation has been reported in the gene encoding for FimH which is a mannose-specific type 1 fimbrial adhesin protein from a range of *Salmonella enterica* serovars. The less binding shear-dependent adhesin is conserved among the wide host range of *Salmonella*. The majority of the FimH variants represent one of the two alternative phenotypes as an appreciably enhanced affinity towards mannose found in *S. typhi* and *S. paratyphi C*, or absolute loss of binding property as in *S. paratyphi B* [15].

3.4 *Shigella*

Annually, *Shigella* spp. causes about 164.7 million cases of shigellosis globally which results in 1.1 million deaths in children of 5-year-old or below. Michigan experienced the largest outbreak in the year 2016 of bacillus dysentery, i.e., shigellosis [16]. The genus *Shigella* comprised of four species *S. boydii, S. sonnei, S. dysenteriae*, and *S. flexneri*. The most frequently isolated species from the patients is *S. flexneri*, mostly prevalent in developing countries [17]. However, *S. sonnei* that causes shigellosis are typically found in developed countries. The efforts to diminish the global burden of shigellosis are a complex process due to low infectious dose and higher antibiotic resistance. In 2017, WHO has announced that *Shigella* should be considered as a priority objective for research and development of therapeutics. *Shigella* is a gram-negative bacterium belongs to the *Enterobacteriaceae* family which also exhibits other enteric pathogens of *E. coli* EPEC, EHEC, EAEC, DAEC, EIEC, and ETEC [18]. The emergence of the various strains of *Shigella* ~25 can be explained by the presence of virulence plasmid. However, the relationship between the plasmid and emergence of species remains controversial for a longer period of time [19]. Advancement of DNA sequencing and phylogenetic analysis has led to the understanding of the clear evolutionary relationship of various *Shigella* spp. along with *E. coli*. The phylogenetic analysis suggests that *Shigella* comprises three major clusters or clades, viz. C1, C2, and C3, with limited numbers of outliers nested to *E. coli*. Furthermore, the C1 has been divided into three subclusters SC1, SC2, and SC3 [20].

3.5 *Pseudomonas aeruginosa*

Similarly, the population of *Pseudomonas aeruginosa* underwent a characteristic evolutionary trend during chronic lung infection described as cystic fibrosis (CF) [21]. The unique conversion has been reported as the reduction of virulence factors, biofilm-associated transition, and higher level of antibiotic resistance [22]. During the CF infection, the population of *P. aeruginosa* exhibits higher phenotypic diversity, toxin production, and resistance which is dynamic over the time imposing a challenging situation of accurate diagnosis and treatment. The spatial heterogeneity and intrinsic spatial structure of CF lung selection seem to play a crucial role in *P. aeruginosa* diversification. A recurrent phenotypic analysis of CF isolates has been found to cause the defects in virulence factors of *P. aeruginosa* such as quorum sensing (QS) and type III secretion [23]. The acquisition of mutation in the QS regulatory genes as LasR causes loss of regulation on QS. Likewise, other mutations include gacS and retS that act as the switch factors between the acute and chronic infection. AmpR has also been considered as the switch factor to chronic infection with a global regulatory network system [24].

3.6 *Helicobacter pylori*

Humans have been colonizing the *Helicobacter pylori* throughout mankind evolution and *H. pylori* has adapted well to the human population. The bacterium colonizes stomach and lives throughout the life. Like other commensal bacteria, this has evolved several mechanisms to evade immune responses. Several strains exhibited factors that modulate the host microenvironment and augmented the risk of gastric adenocarcinoma and peptic ulceration including other diseases. *H. pylori*-encoding genes for these factors have been evolved recombination, mutation, and dynamics of host-pathogen interaction [25]. The most remarkable feature of *H. pylori* is the extreme diversity among the strain primarily in the gene complements, order of the chromosome, exploitation of repetitive DNA, conserved gene sequence variation, mobile DNA, status of phase variable gene, and restriction-modification loci. The variation in the Cag A protein has been seen in various strains of *H. pylori*. The presence or absence of Cag A may define the pathogenicity of the infection. VacA is an exotoxin composed of two subunits, p33 and p55, and almost all the strains of *H. pylori* exhibit *vacA* gene; however only 40% of the strains possess the active form of it [26].

3.7 *Plasmodium falciparum* and vivax

The most important parasitic water-associated infectious disease is malaria which has been evolved along with the human race. On the basis of presence and absence of hemoglobin C homozygous alleles today's population of human malaria parasites *P. falciparum* had originated in West Africa and *P. vivax* in West and Central Africa [27]. Recent molecular evidences have found that humans are the accidental host for malaria which came from apes most probably through the infected mosquito vector bite. The widest global distribution has been attributed to *P. vivax* and *P. malariae*. However, *P. malaria* has lost its predominance and *P. falciparum* including *P. vivax* are the mostly encountered parasites of malaria. Chloroquine (CQ) discovery in 1930s had revolutionized the treatment of malaria and the drug has been most widely used from 1950s to 1990s. After a decade's use mutation in the *P. falciparum* acquired resistance in Thailand and Columbia [28]. Wide genome scanning, discovery of the low variation regions, and signature of selective sweeps allowed us to identify the various genes that underwent mutation throughout the course of evolution. Most importantly mutation in *pfcrt*, *dhfr*, and *dhps* genes causes resistance towards chloroquine, pyrimethamine, and sulfadoxine [29].

3.8 Dengue Virus

Dengue virus (DENV) is a mosquito-borne flavivirus that belongs to family *Flaviviridae*. DENV has been categorized into four serotypes (DENV 1–4). The epidemiological distribution of the serotypes is similar; however, genetically the serotypes are diverse in nature. The phylogenetic analysis has shapes the molecular evolution of DENV serotypes which relies on the complete and partial genomic sequences of all the serotypes and led to the elucidated of the origin, genetic diversity, epidemic potential and transmission dynamics [30, 31]. Using several methods and estimates of nucleotide substitution rate, it has been estimated that DENV-2 genotypes have diverges from the sylvatic forms around 40–600 years ago. However, the emergence of distinct DENV serotypes was most likely influenced by the ecological partitioning and allotropic off-ancestral sylvatic DENV strains in nonhuman primates.

3.9 Conclusions and Future Perspectives

Several factors are responsible for the evolution of pathogens and most imperative are genetic reassortment, mutation, host-pathogen interaction, and environmental causes. Genome-wise analysis is really significant in order to resolve the evolutionary and functional understanding of the pathogens. Due to evolutionary advancement of water-associated pathogens, resistance has been developed against the available therapeutic agents. The pathogens should be categorized in clusters in order to define them phenotypically and genetically. In order to design the effective vaccine strategies one should focus on the consensus sequence pattern among all the species.

References

1. Pandey PK, Kass PH, Soupir ML, Biswas S, Singh VP (2014) Contamination of water resources by pathogenic bacteria. AMB Express 4:51
2. Kummu M, Guillaume JH, de Moel H, Eisner S, Flörke M, Porkka M, Siebert S, Veldkamp TI, Ward PJ (2016) The world's road to water scarcity: shortage and stress in the 20th century and pathways towards sustainability. Sci Rep 6:38495
3. Conway DJ, Roper C (2000) Micro-evolution and emergence of pathogens. Int J Parasitol 30(12–13):1423–1430
4. Metcalf CJ, Birger RB, Funk S, Kouyos RD, Lloyd-Smith JO, Jansen VA (2015) Five challenges in evolution and infectious diseases. Epidemics 10:40–44
5. Rahaman MH, Islam T, Colwell RR, Alam M (2015) Molecular tools in understanding the evolution of Vibrio cholerae. Front Microbiol 6:1040
6. Greig DR, Schaefer U, Octavia S, Hunter E, Chattaway MA, Dallman TJ, Jenkins C (2018) Evaluation of whole-genome sequencing for identification and typing of Vibrio cholerae. J Clin Microbiol 56(11):e00831-18
7. Keddy KH, Nadan S, Govind C, Sturm AW, Group for Enteric, Respiratory and Meningeal Disease Surveillance in South Africa (2007) Evidence for a clonally different origin of the two cholera epidemics of 2001–2002 and 1980–1987 in South Africa. J Med Microbiol 56(Pt 12):1644–1650
8. Pradhan S, Baidya AK, Ghosh A, Paul K, Chowdhury R (2010) The El Tor biotype of Vibrio cholerae exhibits a growth advantage in the stationary phase in mixed cultures with the classical biotype. J Bacteriol 192(4):955–963

9. Chattopadhyay DJ, Sarkar BL, Ansari MQ, Chakrabarti BK, Roy MK, Ghosh AN, Pal SC (1993) New phage typing scheme for Vibrio cholerae O1 biotype El Tor strains. J Clin Microbiol 31(6):1579–1585
10. Kitaoka M, Miyata ST, Unterweger D, Pukatzki S (2011) Antibiotic resistance mechanisms of Vibrio cholerae. J Med Microbiol 60(Pt 4):397–407
11. Faruque SM, Sack DA, Sack RB, Colwell RR, Takeda Y, Nair GB (2003) Emergence and evolution of Vibrio cholerae O139. Proc Natl Acad Sci U S A 100(3):1304–1309
12. Kisiela DI, Chattopadhyay S, Libby SJ, Karlinsey JE, Fang FC, Tchesnokova V, Kramer JJ, Beskhlebnaya V, Samadpour M, Grzymajlo K, Ugorski M, Lankau EW, Mackie RI, Clegg S, Sokurenko EV (2012) Evolution of Salmonella enterica virulence via point mutations in the fimbrial adhesin. PLoS Pathog 8(6):e1002733
13. Umann D, Cunrath O (2017) Heterogeneity of Salmonella-host interactions in infected host tissues. Curr Opin Microbiol 39:57–63
14. Moreno Switt AI, den Bakker HC, Cummings CA, Rodriguez-Rivera LD, Govoni G, Raneiri ML, Degoricija L, Brown S, Hoelzer K, Peters JE, Bolchacova E, Furtado MR, Wiedmann M (2012) Identification and characterization of novel Salmonella mobile elements involved in the dissemination of genes linked to virulence and transmission. PLoS One 7(7):e41247
15. Ido N, Lee K, Iwabuchi K, Izumiya H, Uchida I, Kusumoto M, Iwata T, Ohnishi M, Akiba M (2014) Characteristics of Salmonella enterica serovar 4,[5],12:i:- as a monophasic variant of serovar Typhimurium. PLoS One 9(8):e104380
16. Doore SM, Schrad JR, Dean WF, Dover JA, Parent KN (2018) Shigella phages isolated during a dysentery outbreak reveal uncommon structures and broad species diversity. J Virol 92(8):e02117-17
17. Zhao L, Xiong Y, Meng D, Guo J, Li Y, Liang L, Han R, Wang Y, Guo X, Wang R, Zhang L, Gao L, Wang J (2017) An 11-year study of shigellosis and Shigella species in Taiyuan, China: active surveillance, epidemic characteristics, and molecular serotyping. J Infect Public Health 10(6):794–798
18. Gomes TA, Elias WP, Scaletsky IC, Guth BE, Rodrigues JF, Piazza RM, Ferreira LC, Martinez MB (2016) Diarrheagenic Escherichia coli. Braz J Microbiol 47(Suppl 1):3–30
19. Sayeed S, Reaves L, Radnedge L, Austin S (2000) The stability region of the large virulence plasmid of Shigella flexneri encodes an efficient postsegregational killing system. J Bacteriol 182(9):2416–2421
20. Yang J, Nie H, Chen L, Zhang X, Yang F, Xu X, Zhu Y, Yu J, Jin Q (2007) Revisiting the molecular evolutionary history of Shigella spp. J Mol Evol 64(1):71–79
21. Winstanley C, O'Brien S, Brockhurst MA (2016) Pseudomonas aeruginosa evolutionary adaptation and diversification in cystic fibrosis chronic lung infections. Trends Microbiol 24(5):327–337
22. Hosseinidoust Z, van de Ven TG, Tufenkji N (2013) Evolution of Pseudomonas aeruginosa virulence as a result of phage predation. Appl Environ Microbiol 79(19):6110–6116
23. Sadikot RT, Blackwell TS, Christman JW, Prince AS (2005) Pathogen-host interactions in Pseudomonas aeruginosa pneumonia. Am J Respir Crit Care Med 171(11):1209–1223
24. Moradali MF, Ghods S, Rehm BH (2017) Pseudomonas aeruginosa lifestyle: a paradigm for adaptation, survival, and persistence. Front Cell Infect Microbiol 15(7):39
25. Haley KP, Gaddy JA (2015) Helicobacter pylori: genomic Insight into the host-pathogen interaction. Int J Genomics 2015:386905
26. Jones KR, Whitmire JM, Merrell DS (2010) A tale of two toxins: Helicobacter pylori CagA and VacA modulate host pathways that impact disease. Front Microbiol 1:115
27. Loy DE, Liu W, Li Y, Learn GH, Plenderleith LJ, Sundararaman SA, Sharp PM, Hahn BH (2017) Out of Africa: origins and evolution of the human malaria parasites Plasmodium falciparum and Plasmodium vivax. Int J Parasitol 47(2–3):87–97
28. Baird JK (2009) Resistance to therapies for infection by Plasmodium vivax. Clin Microbiol Rev 22(3):508–534

29. Sridaran S, Rodriguez B, Soto AM, Macedo De Oliveira A, Udhayakumar V (2014) Molecular analysis of chloroquine and sulfadoxine-pyrimethamine resistance-associated alleles in Plasmodium falciparum isolates from Nicaragua. Am J Trop Med Hyg 90(5):840–845
30. Sessions OM, Wilm A, Kamaraj US, Choy MM, Chow A, Chong Y, Ong XM, Nagarajan N, Cook AR, Ooi EE (2015) Analysis of dengue virus genetic diversity during human and mosquito infection reveals genetic constraints. PLoS Negl Trop Dis 9(9):e0004044
31. Hanley KA, Guerbois M, Kautz TF, Brown M, Whitehead SS, Weaver SC, Vasilakis N, Marx PA (2014) Infection dynamics of sylvatic dengue virus in a natural primate host, the African Green Monkey. Am J Trop Med Hyg. 91(4):672–676

Epidemiology of Water-Associated Infectious Diseases

4

Swatantra Kumar*, Amrita Haikerwal,
and Shailendra K. Saxena*

Abstract

Infection pervasiveness is significantly related to the exposure and rate of transmission which are influenced by ecological factors, for example, precipitation, air/water temperature, and seasonal variability. *Vibrio cholerae* is solely responsible for approximately 1.7 million cases annually with 525,000 deaths in children below 5 years. Similarly, enteric fever (typhoid) is a severe systemic infection and is the foremost public health water-borne infectious disease with an estimated 26 million cases annually in the same way. *Giardia intestinalis* is the foremost cause of parasitic infection in the USA with an estimated 1.2 million cases and 3581 reported hospitalizations annually. So far, three species of schistosome have been archived including *Schistosoma haematobium* which causes urogenital disease in sub-Saharan Africa. According to WHO-World Malaria Report-2016, 212 million cases along with 429,000 deaths were reported in the year 2015. Shigellosis is caused by a group of bacteria known as *Shigella* with an estimated 500,000 cases annually in the USA. Recently around 6000 cases of Legionellosis were reported in the USA in the year 2015.

Keywords

Transmission risk · Case fatality rate · Water-borne · Global technical strategy
Life cycle · Prevalence · Morbidity · Mortality

---

*Contributed equally as first author

S. Kumar · A. Haikerwal · S. K. Saxena (✉)
Centre for Advanced Research, Faculty of Medicine, King George's Medical University, Lucknow, India
e-mail: shailen@kgmcindia.edu

S. K. Saxena (ed.), *Water-Associated Infectious Diseases*,
https://doi.org/10.1007/978-981-13-9197-2_4

4.1 Introduction

The relation between various domains of water-associated pathogens and epidemiology (Fig. 4.1) has been poorly established and can be effectively demonstrated by various hydro-climatological assessments [1]. The dynamics of infection and water-associated infectious disease load are mainly related to hydro-climatological drivers across a broad range of temporal and spatial regions [2]. Infection commonness can be altogether corresponded with the exposure and transmission rate that are influenced by natural factors, for example, precipitation, seasonal variability, and air/water temperature [3]. For example, a flood in a territory may result in the tainting of drinking water sources with crude sewage. This may lead to the washing out of the etiological agents from open defecation sites [4]. As a rule precipitation may influence human exercises identified with individual cleanliness and water contacts possibly expanding the transmission hazard [5]. Another imperative factor may cause the seasonal variability in disease transmission caused by the temporal fluctuation of the rainfall [6]. Recreational settings and drinking water are the well-known sources of water-associated infections. Some of them are highlighted in Table 4.1.

One of the leading causes of mortality associated with water is diarrhea [7] caused by *Vibrio cholerae*. This water-borne infectious disease is solely responsible for approximately 1.7 million cases annually with 525,000 deaths in children below 5 years [8]. Similarly enteric fever (typhoid) is a severe systemic infection for public health and the foremost water-borne infectious disease with 26 million annual cases [9]. Typhoid is stimulated by ingestion of the *Salmonella enterica* serotypes which

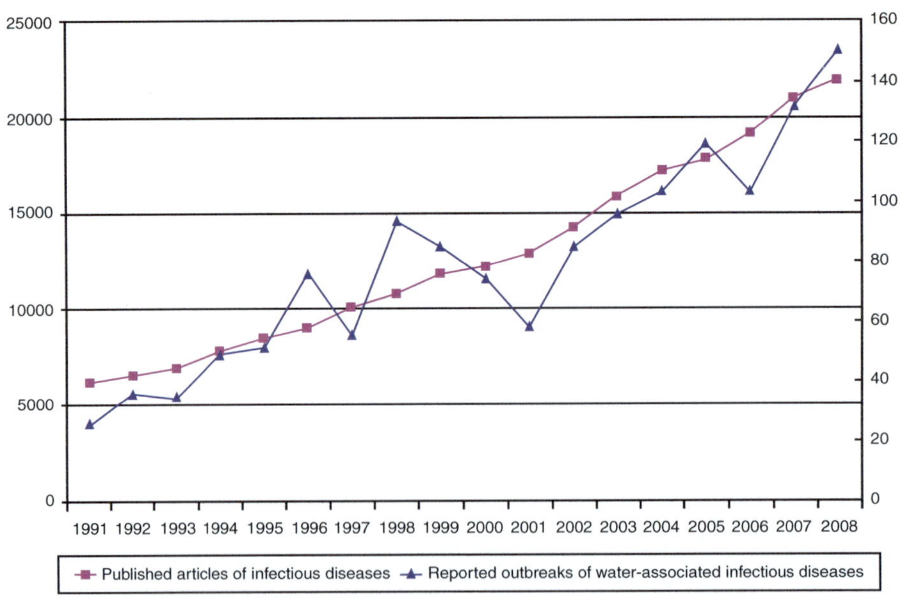

Fig. 4.1 Trend of articles published in the area of infectious diseases and epidemics [adapted from PLoS Negl Trop Dis 6(2): e1483]

Table 4.1 Water-associated pathogens, diseases, and primary source of infection

Diseases	Reservoir and primary sources	Pathogen
Enteropathogenic bacteria		
Typhoid fever	Human feces	*Salmonella typhi*
Paratyphoid fever	Human feces	*Salmonella paratyphi*
Salmonellosis	Human and animal feces	Other *Salmonella*
Bacillary dysentery	Human feces	*Shigella* spp.
Cholera	Human feces and freshwater zooplankton	*Vibrio cholera*
Gastroenteritis	Human feces	Enteropathogenic *E. coli*
Hemorrhagic colitis	Animal feces	*Escherichia coli* O157:H7 (VETC)
Gastroenteritis	Human and animal feces	*Yersinia enterocolitica*
Gastroenteritis	Human and animal feces	*Campylobacter jejuni*
Putative bacterial pathogens grown in water supply		
Shanghai fever	Surface water contaminated with effluents	*Pseudomonas aeruginosa*
Diarrhea illness		*Aeromonas*
Legionellosis	Thermally enriched water	*Legionella pneumophila* and related bacteria
Leptospirosis	Animal and human urine	*Leptospira* spp.
Pulmonary illness	Soil and water, recreational settings	Nontuberculosis Mycobacteria (NTMs)
Chronic gastritis, peptic ulcer disease, or gastric cancer	Human feces	*Helicobacter pylori*
Variable	Natural water	Opportunistic bacteria
Enteric viruses		
Poliomyelitis	Human feces	Polio viruses
Aseptic meningitis	Human feces	Coxsackie viruses A
Aseptic meningitis	Human feces	Coxsackie viruses B
Aseptic meningitis	Human feces	Echo viruses
Encephalitis	Human feces	Other enteroviruses
Gastroenteritis	Human feces	Rotaviruses
Upper respiratory and gastrointestinal illness	Human feces	Adenoviruses
Infectious hepatitis	Human feces	Hepatitis A virus
Infectious hepatitis; miscarriage and death	Human feces	Hepatitis E virus
Gastroenteritis	Fomites and water	Norovirus
Protozoa		
Amoebic meningoencephalitis	Human feces	*Acanthamoeba castellanii*
Balantidiasis (dysentery)	Human and animal feces	*Balantidium coli*
Cryptosporidiosis (gastroenteritis)	Water, human, and other mammal feces	*Cryptosporidium hominis, C. parvum*
Amoebic dysentery	Human and animal feces	*Entamoeba histolytica*
Giardiasis (gastroenteritis)	Water and animal feces	*Giardia lamblia*
Primary amoebic meningoencephalitis	Warm water	*Naegleria fowleri*

(continued)

Table 4.1 (continued)

Diseases	Reservoir and primary sources	Pathogen
Helminths		
Ascariasis	Human and animal feces	*Ascarislum bricoides*
Other water-related pathogens		
Malaria	*Anopheles gambiae*	*Plasmodium falciparum, P. vivax*
Sleeping sickness	Tsetse flies	*Trypanosoma brucei*
Lymphatic filariasis	Culex mosquito	*Wuchereria bancrofti*
Encephalitis	*Culex tritaeniorhynchus*	Japanese encephalitis virus
Dengue hemorrhagic fever	*Aedes aegypti*	Dengue virus
Microcephaly, Guillain-Barre syndrome	*Aedes aegypti, Aedes albopictus*	Zika virus
Encephalitis	*Aedes albopictus, Aedes aegypti*, and *Culex tritaeniorhynchus*	West Nile virus
Yellow fever	*Aedes aegypti*	Yellow fever virus

are typhi and paratyphi (A, B, and C) where the infection is known to restrict in humans only [10]. Water-carried etiological agents are a part of water-borne pathogens and are explained as the unintentional ingestion of contaminated water in recreational settings [11]. According to the Morbidity and Mortality Weekly Report (MMWR), CDC, 52% of the outbreaks in the recreation setting is caused by *Cryptosporidium*, a parasite which can survive in the chlorinated water (chlorine-tolerant parasite) for more than 10 days. Similarly, *Giardia intestinalis* is the foremost cause of parasitic infection in the USA with approximately 1.2 million cases and 3581 reported hospitalizations annually [12]. Several routes of Giardiasis infection have been documented but the most effective is water-borne (74.6%) which is related to drinking water and recreational settings (primarily swimming pools) [13].

Dissimilar to water-borne, water-based causative agents are primarily worms that spend a fraction of their life cycle in water bodies. Schistosomiasis is caused by parasitic worms known as *cercariae* in humans through contact with water. Currently, approximately 206 million people are infected with larva and approximately 700 million people are susceptible [14]. Recently, three schistosome species have been recognized including *Schistosoma haematobium* that are responsible for urogenital infection in sub-Saharan Africa [15] and *Schistosoma japonicum* for intestinal infection in the Philippines, Indonesia [16], and People's Republic of China [17]. Similarly, *S. mansoni* is a well-known intestinal infection in sub-Saharan Africa, Caribbean island, and Brazil [18]. Water-related diseases are characterized as vector-borne infectious diseases where malaria and trypanosomiasis are most significant [19]. Epidemiology of the malaria can be described as highly heterogeneous and focal with parasites and vector diversity, seasonal variability, and transmission intensity [20]. As per World Malaria report-2016, WHO, there were 212 million cases with 429,000 deaths in the year 2015 [21]. Global technical strategy (GTS) for malaria aims to reduce the global burden of malaria by 2030 by at least

90% as compared with cases reported during 2015 [22]. International Centers of Excellence of Malaria Research (ICEMR) has created the epidemiological settings comprising endemic regions for *P. vivax* or *P. falciparum* to obligate surveillance [23]. Similarly, sleeping sickness or human African trypanosomiasis (HAT) is caused by protozoa identified as trypanosomes. The infection transmission is pervasive in sub-Saharan African nations where it is known to spread via a fly's tsetse [24]. From most recent 50 years, significant decrease in the number of emerging cases has been accomplished in the year 2009 with the number of cases below 10,000 [25]. Furthermore, water-washed diseases may appear due to poor individual and residential cleanliness due to the contact with contaminated water and clean water deprivation. Shigellosis is caused by a set of bacteria known as *Shigella* with an estimated 500,000 cases in the USA annually [26]. Shigellosis results in fever, diarrhea, and stomach cramps and being highly contagious its source of infection can be diverse as skin contact, contaminated food, and swallowing recreational water while swimming. Similarly, trachoma is caused by a bacterium *Chlamydia trachomatis* which easily spreads through personal contact and causes trichiasis [27]. The disease is typically widespread in regions with inadequately developed water hygiene facility with almost 8 million people visually harmed and with about 500 million at the risk of infection [28].

Likewise, Pontiac fever (Legionellosis) is the well-known water-dispersed disease caused by *Legionella*. Legionella outbreak is commonly associated with buildings having complex water system supply [29]. Recently around 6000 cases of Legionellosis were reported in the USA in the year 2015 [30]. It has been estimated that 1 in every 10 person who gets diagnosed with Legionellosis will die because of respiratory failure [31] (Fig. 4.2). Water-associated pathogens, diseases, and primary source of infection are given in Table 4.1 [32].

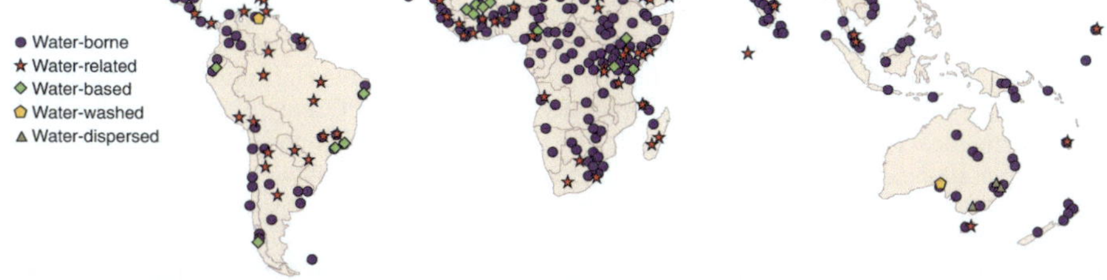

Fig. 4.2 Reported outbreak distribution of WAID from 1991 to 2008 [adapted from PLoS Negl Trop Dis 6(2): e1483. doi: https://doi.org/10.1371/journal.pntd.0001483]

References

1. Mari L, Casagrandi R, Bertuzzo E, Rinaldo A, Gatto M (2014) Floquet theory for seasonal environmental forcing of spatially explicit waterborne epidemics. Theor Ecol 7(4):351–365
2. Burnet JB, Penny C, Ogorzaly L, Cauchie HM (2014) Spatial and temporal distribution of Cryptosporidium and Giardia in a drinking water resource: implications for monitoring and risk assessment. Sci Total Environ 472:1023–1035
3. Cann KF, Thomas DR, Salmon RL, Wyn-Jones AP, Kay D (2013) Extreme water-related weather events and waterborne disease. Epidemiol Infect 141(4):671–686
4. de Man H, van den Berg HH, Leenen EJ, Schijven JF, Schets FM, van der Vliet JC, van Knapen F, de Roda Husman AM (2014) Quantitative assessment of infection risk from exposure to waterborne pathogens in urban floodwater. Water Res 48(Jan.):90–99
5. Carlton EJ, Eisenberg JN, Goldstick J, Cevallos W, Trostle J, Levy K (2014) Heavy rainfall events and diarrhea incidence: the role of social and environmental factors. Am J Epidemiol 179(3):344–352
6. Etchie AT, Etchie TO, Adewuyi GO, Kannan K, Wate SR, Sivanesan S, Chukwu AU (2014) Influence of seasonal variation on water quality in tropical water distribution system: is the disease burden significant? Water Res 49:186–196
7. Lanata CF, Fischer-Walker CL, Olascoaga AC, Torres CX, Aryee MJ, Black RE (2013) Child Health Epidemiology Reference Group of the World Health Organization and UNICEF. Global causes of diarrheal disease mortality in children <5 years of age: a systematic review. PLoS One 8(9):e72788
8. World Health Organization (2017) Diarrhoeal disease fact sheet. WHO, Geneva. http://www.who.int/mediacentre/factsheets/fs330/en/. Accessed on 12-3-2018
9. Judd MC, Mintz ED (2017) Typhoid & paratyphoid fever. Centers for Disease Control and Prevention. National Center for Emerging and Zoonotic Infectious Diseases (NCEZID). Division of Global Migration and Quarantine (DGMQ), 31 May 2017. https://wwwnc.cdc.gov/travel/yellowbook/2018/infectious-diseases-related-to-travel/typhoid-paratyphoid-fever. Accessed on 12-3-2018.
10. John J, van Aart CJ, Grassly NC (2016) The burden of typhoid and paratyphoid in India: systematic review and meta-analysis. PLoS Negl Trop Dis 10(4):e0004616
11. Conn DB (2014) Aquatic invasive species and emerging infectious disease threats: a one health perspective. Aquat Invasions 9(3):383–390
12. Painter JE et al (2015) Centers for Disease Control and Prevention. Cryptosporidiosis surveillance—United States, 2011–2012. Giardiasis surveillance—United States, 2011–2012. MMWR Suppl 64(3):1–25
13. Strathmann M, Horstkott M, Koch C, Gayer U, Wingender J (2016) The River Ruhr—an urban river under particular interest for recreational use and as a raw water source for drinking water: the collaborative research project "Safe Ruhr"—microbiological aspects. Int J Hyg Environ Health 219(7 Pt B):643–661
14. World Health Organization (2018) Schistosomiasis fact sheet. WHO, Geneva. http://www.who.int/mediacentre/factsheets/fs115/en/. Accessed on 12-3-2018
15. Brindley PJ, Hotez PJ (2013) Break out: urogenital schistosomiasis and Schistosoma haematobium infection in the post-genomic era. PLoS Negl Trop Dis 7(3):e1961
16. Soares Magalhães RJ, Salamat MS, Leonardo L, Gray DJ, Carabin H, Halton K, McManus DP, Williams GM, Rivera P, Saniel O, Hernandez L, Yakob L, McGarvey S, Clements A (2014) Geographical distribution of human Schistosoma japonicum infection in the Philippines: tools to support disease control and further elimination. Int J Parasitol 44(13):977–984
17. Chen MG (2014) Assessment of morbidity due to Schistosoma japonicum infection in China. Infect Dis Poverty 3(1):1–16
18. Nascimento LS, Ebert JP, França VT, Cruz DB (2016) Update on Schistosomiasis mansoni: epidemiology, pathogenesis, diagnosis, treatment and prevention. JSM Gastroenterol Hepatol 4(4):1068

19. Campbell OM, Benova L, Gon G, Afsana K, Cumming O (2015) Getting the basic rights—the role of water, sanitation and hygiene in maternal and reproductive health: a conceptual framework. Tropical Med Int Health 20(3):252–267
20. Eckhoff P (2013) Mathematical models of within-host and transmission dynamics to determine effects of malaria interventions in a variety of transmission settings. Am J Trop Med Hyg 88(5):817–827
21. World Health Organization (2016) World malaria report. WHO, Geneva. http://apps.who.int/iris/bitstream/10665/252038/1/9789241511711-eng.pdf?ua=1. Accessed on 12-3-2018
22. Patouillard E, Griffin J, Bhatt S, Ghani A, Cibulskis R (2017) Global investment targets for malaria control and elimination between 2016 and 2030. BMJ Glob Health 2(2):e000176
23. Moss WJ, Dorsey G, Mueller I, Laufer MK, Krogstad DJ, Vinetz JM, Guzman M, Rosas-Aguirre AM, Herrera S, Arevalo-Herrera M, Chery L, Kumar A, Mohapatra PK, Ramanathapuram L, Srivastava HC, Cui L, Zhou G, Parker DM, Nankabirwa J, Kazura JW, International Centers of Excellence for Malaria Research (2015) Malaria epidemiology and control within the International Centers of Excellence for Malaria Research. Am J Trop Med Hyg. 93(3 Suppl):5–15
24. Franco JR, Simarro PP, Diarra A, Jannin JG (2014) Epidemiology of human African trypano somiasis. Clin Epidemiol 6:257–275
25. World Health Organization (2014) Human African trypanosomiasis: epidemiological situation. WHO, Geneva. http://www.who.int/trypanosomiasis_african/country/en/. Accessed on 12-3-2018.
26. Centers for Disease Control and Prevention (2017) National Center for Emerging and Zoonotic Infectious Diseases (NCEZID). Division of Global Migration and Quarantine (DGMQ). Shigella—Shigellosis. CDC, Atlanta, GA. https://www.cdc.gov/shigella/general-information.html. Accessed on 12-3-2018
27. George N, Levine AC (2016) Ocular Chlamydia trachoma. Can J Emerg Med 18(5):401–406
28. Centers for Disease Control and Prevention (2009) National Center for Emerging and Zoonotic Infectious Diseases (NCEZID). Division of Global Migration and Quarantine (DGMQ). Water, sanitation & environmentally-related hygiene. Hygiene-related diseases. CDC, Atlanta, GA. https://www.cdc.gov/healthywater/hygiene/disease/trachoma.html#one. Accessed on 12-03-2018
29. Centers for Disease Control and Prevention (2017) National Center for Immunization and Respiratory Diseases, Division of Bacterial Diseases. Legionella (Legionnaires' disease and Pontiac Fever). CDC, Atlanta, GA. https://www.cdc.gov/legionella/outbreaks.html. Accessed on 12-3-2018
30. Centers for Disease Control and Prevention (2017) National Center for Immunization and Respiratory Diseases, Division of Bacterial Diseases. Legionella (legionnaires' disease and Pontiac Fever), Fast facts. CDC, Atlanta, GA. https://www.cdc.gov/legionella/fastfacts.html#f01. Accessed on 12-3-2018
31. Correia AM, Ferreira JS, Borges V, Nunes A, Gomes B, Capucho R, Gonçalves J, Antunes DM, Almeida S, Mendes A, Guerreiro M, Sampaio DA, Vieira L, Machado J, Simões MJ, Gonçalves P, Gomes JP (2016) Probable person-to-person transmission of legionnaires' disease. N Engl J Med 374(5):497–498
32. Saxena SK, Kumar S, Haikerwal A, Bhatt ML (2018) Introductory chapter: Neglected tropical waterborne infectious diseases-strategies for mitigation. In: Water challenges of an urbanizing world. IntechOpen, London

Emerging and Re-emerging Water-Associated Infectious Diseases

5

Swatantra Kumar*, Vimal K. Maurya, and Shailendra K. Saxena*

Abstract

Expanding human population, globalization, deforestation, altered ecosystem, and microbial evolution are the most convergent forces for the evolution of infectious diseases which are associated with water. Emerging and re-emerging water-associated infectious diseases are defined as infectious diseases which have newly emerged among a population or existing in nature with rapid expansion of incident or geographic range. In order to overcome these infections, better prevention strategies such as an effective vaccination and advancement in drug discovery are crucial. Various newly or emerging pathogens have been evolved due to contaminated water in the recent years. The factors for the arising of novel pathogens are the escalating number of human population vulnerable to specific probable pathogens. Changes in human demographics and behavior, breakdown of public health measures, microbial adaptation, and changes in agricultural practices are the major factors responsible for the progression of water-associated infectious diseases.

Keywords

Typhoid · Legionellosis · Leptospirosis · *Helicobacter pylori* · Coxsackie viruses · Rotaviruses · Dengue · Zika

*Contributed equally as first author

S. Kumar · V. K. Maurya · S. K. Saxena (✉)
Centre for Advanced Research, Faculty of Medicine, King George's Medical University, Lucknow, India
e-mail: shailen@kgmcindia.edu

S. K. Saxena (ed.), *Water-Associated Infectious Diseases*,
https://doi.org/10.1007/978-981-13-9197-2_5

5.1 Introduction

Emerging and re-emerging water-associated infectious diseases always have been a global threat to the mankind in conjunction with innumerable therapeutic challenges to the scientific community. Emerging infectious diseases have recently appeared within a population and it has a potential to spread in future. Expanding human population, globalization, deforestation, altered ecosystem, and microbial evolution are the most convergent forces for the emergence of water-associated infectious diseases [1]. Emerging and re-emerging water-associated infectious diseases are defined as diseases which have newly appeared among a population or existing in nature with rapid expansion of incident or geographic range. These infectious diseases are mostly water-associated infectious diseases including bacteria, fungi, parasites, and viruses which have been a global threat for the mankind. Typhoid, bacillus dysentery, cholera, enteropathogenic *E. coli*, *Yersinia enterocolitica*, *Campylobacter jejuni*, *Pseudomonas aeruginosa*, *Aeromonas*, legionellosis, leptospirosis, *Helicobacter pylori*, polio virus, Coxsackie virus A, Coxsackie virus B, rotaviruses, adenoviruses, hepatitis A virus, norovirus, *Cryptosporidium hominis*, *Entamoeba histolytica*, *Giardia lamblia*, *Naegleria fowleri*, *Ascaris lumbricoides*, malaria, filaria, Japanese encephalitis, dengue, Zika, West Nile, and yellow fever have been discussed here. Disease symptoms, transmission, and life cycle of all these have been discussed in this chapter. In order to overcome these infections, improved prevention strategies such as an effective vaccination and advancement in drug discovery are crucial [2]. In the modern era, various new or emerging pathogens have been evolved due to contaminated water. Therefore it is imperative to attribute the factors that can cause emergence of water-associated infectious diseases.

5.2 Alteration in Human Demographics and Behavior

Evolutionary factors for novel pathogens are defined as the rising number of human population vulnerable to specific probable pathogens. This susceptible population includes immunocompromised state such as in case of HIV and patients receiving immunosuppressive drugs for various causes like cancer and organ transplantation. As a result of reduced level of immunity these vulnerable populations are susceptible to various infections that are not reported in healthy individuals. Other prevalent populations which are prone to emerging infections are young or newborn babies due to underdeveloped immune system or due to low socioeconomic profile. Rural urbanization allows infections to emerge in the isolated areas which may remain obscure or localized to affect larger populations. Moreover, the worldwide growth in urban population has resulted in the poor hygiene practices such as waste or contaminated water [3].

5.3 Failure of Public Health Measures

The classical public health procedures have been continuously serving as to minimize the human exposure to several pathogens. However, some pathogen remains in the host reservoirs in inactive stage or in the environment and they often take the advantage of emergence if the public health measures break down. One of the promising examples of this is cholera outbreak in South America due to reduction of chlorine level in the water supplies. Similarly, the cryptosporidium infection has emerged in Wisconsin due to nonfunctioning filtration system in water supplies. Limitation of surveillance and lack of availability of diagnostic procedure constraint public health measures to control the emergence of water-associated infectious diseases [4].

5.4 Microbial Adaptation

The evolution of the microorganisms may change the virulence and pathogenic properties of water-associated infectious diseases. The most important example of this is pathogenic *Escherichia coli* strains which have acquired the virulent gene via horizontal gene transfer resulting in the highly pathogenic strain as EHEC (enterohemorrhagic *E. coli*). Emergence of antibiotic resistance bacteria and intense use of antibiotics in clinical settings have led to the development and progression of antibiotic resistance infectious agents. Nonpathogenic species may acquire the resistance gene from the other species and this results in the emergence of deadly strains against humans [5].

5.5 Changes in Agricultural Practices

Agricultural production methods may also result in the emergence of the novel pathogens. As the quantity of livestock farms is declining, agricultural farms are merging into intensive farming activities resulting in the concentration of animal waste, and increase in the pollution of natural water resources and streams by agricultural runoff and waste. Drinking water supplies are the major source of the emergence of any water-associated pathogen in a locality which might be due to contamination by animal waste [6].

5.6 Emerging and Re-emerging Water-Associated Infectious Diseases

5.6.1 Typhoid

Typhoid fever is caused by *Salmonella* species and usually has a persistent fever as high as 103–104 °F (39–40 °C). Peoples infected with Salmonella may also feel

Fig. 5.1 Typhoid fever symptoms

stomach pain, weakness, loss of appetite, and headache (Fig. 5.1) [7]. Immense use of antibiotics for the management of typhoid is responsible for the emergence of multidrug resistance *Salmonella* species [8].

5.6.2 Bacillus Dysentery

Dysentery is a chronic form of diarrhea and also known as shigellosis with excessive blood and mucus in the stools. Bacillary dysentery is related with species of bacteria from the family *Enterobacteriaceae*. Colic pains and frequent urging are the two characterized symptoms in severe dysentery. The onset is typically rapid, accompanied by a high fever, thirst, dehydration with fatigue, and depletion of strength (Fig. 5.2) [9].

5.6.3 Cholera

Cholera is a contagious disease and mainly caused by drinking water or eating food contaminated with a toxigenic bacterium called *Vibrio cholerae*. Patients infected with *Vibrio cholerae* is characterized by profuse watery diarrhea, leg cramps, vomiting, dehydration, and shock. Cholera is a worldwide risk to public health and an indicator of inequality and social development deficit. Cholera can be mild or even without symptoms (Fig. 5.3), but extreme cases can lead to death if untreated [10].

Fig. 5.2 Steps towards protection from Shigellosis

The acute cases of typhoid are mainly characterized by prolonged fever, loss of appetite, headache, nausea, and constipation or sometimes diarrhea [11].

5.6.4 Enteropathogenic *E. coli*

Enteropathogenic *Escherichia coli* (EPEC) is a gram-negative bacterium and causes diarrhea in humans and animals (Fig. 5.4) [12]. It has been a major risk to human health and remains as a major cause of newborn child's deaths in developing countries. Although EPEC was the primary *E. coli strain* to be occupied in human disease in the 1940s and 1950s, the mechanisms by which this pathogen induced diarrhea remained unknown through the vast majority of a long time since its depiction.

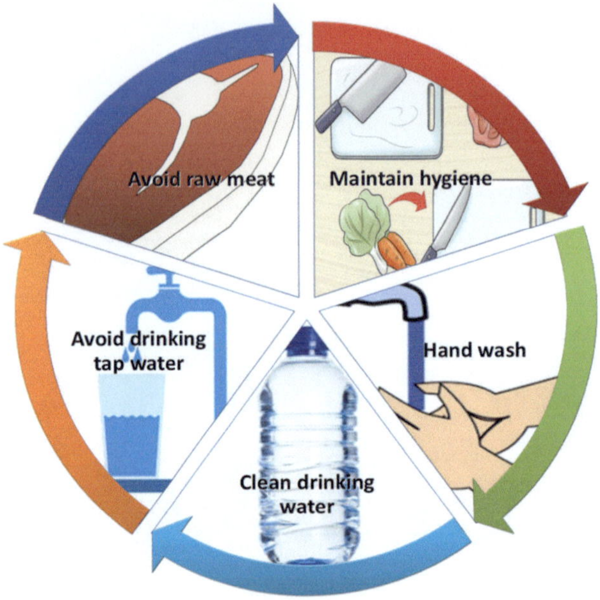

Fig. 5.3 Public health measures for the prevention of cholera

Fig. 5.4 Symptoms during EPEC infection

5.6.5 *Yersinia enterocolitica*

Yersiniosis is a typical zoonotic foodborne infection. *Yersinia enterocolitica* are the bacterial species of family *Enterobacteriaceae* and are responsible for various diseases in humans including enterocolitis, lymphadenitis, acute diarrhea, terminal ileitis, and mesenteric and pseudo appendicitis. The systematic infection of *Yersinia*

Fig. 5.5 Symptoms during *Campylobacter jejuni* infection

enterocolitica may result in fatal sepsis [13]. Symptoms like watery feces diarrhea and weight loss with its functional correlates diminish transepithelial electrical resistance and enhance permeability of fluorescein. Risk factors for securing of the microorganisms for newborn children, who are more regularly influenced than adults, were distinguished as playing in a sandbox or having contact with pets.

5.6.6 *Campylobacter jejuni*

The most common bacterial infections in humans are due to *Campylobacter* infections. It produces both systemic illnesses and diarrheal symptoms (Fig. 5.5). In developed regions, enteric *Campylobacter* infections produce an inflammatory, sometimes bloody, diarrhea [14]. Complications include toxic megacolon, dehydration, and sepsis in children and immunocompromised patients.

5.6.7 *Pseudomonas aeruginosa*

Pseudomonas aeruginosa is known as one of the most crucial opportunistic pathogens in humans. Infected patients with compromised immune systems such as immunodeficiency or burns, cystic fibrosis, and hematological malignancy are at greater risk of getting infection. Enteric disease associated with sepsis caused by *Pseudomonas aeruginosa* has been reported as early as 1918 with the story of "Shanghai fever," a syndrome comprising fever, diarrhea, and sepsis [15].

Fig. 5.6 Aeromonas infection

5.6.8 *Aeromonas*

Aeromonas are known to cause a wide spectrum of diseases. Holmberg and colleagues first reported the ecological source of *Aeromonas* causing gastrointestinal infection. It was presented as an epidemiological research to sustain the importance of crude well water in patients with diarrheal disease as a source of infection (Fig. 5.6) [16].

5.6.9 Legionellosis

Symptoms of Legionnaires' infection usually develop faster, including high fever, dry cough, and shortness of breath, often accompanied by headache, muscle pain, tiredness, gastrointestinal problems (Fig. 5.7), and mental confusion [17]. There is no vaccine available for the treatment of legionellosis.

5.6.10 Leptospirosis

Leptospira spp. pathogenic strains are gram-negative bacteria which cause a serious infectious disease called leptospirosis. Generally, leptospirosis is known as an occupational disease that affects people residing in rural areas, primarily farmers involved in agriculture or breeding of animals considered to be the major groups at risk (Fig. 5.8) [18]. Leptospirosis can lead to meningitis, kidney damage, respiratory distress, liver failure, and even death if it remains untreated.

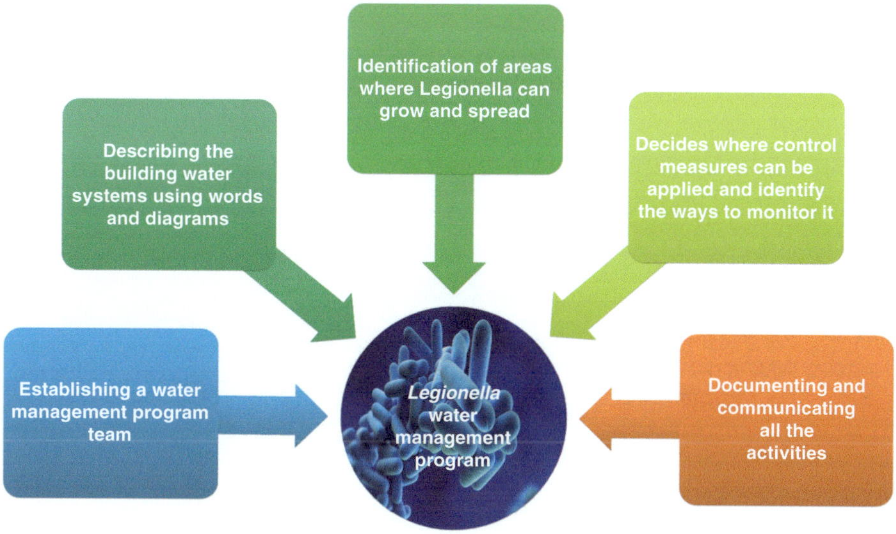

Fig. 5.7 Legionnaire's water management system

Fig. 5.8 Symptoms of Leptospirosis

5.6.11 *Helicobacter pylori*

Helicobacter pylorus (*H. pylori*) causes stomach infection and is a leading source of gastritis and peptic ulcers (Fig. 5.9). Chronic *H. pylori* infection to the stomach may lead to the stomach cancer. Contaminated water and food are the primary source of infection in humans [19]. The presence of *H. pylori* in the upper gastrointestinal

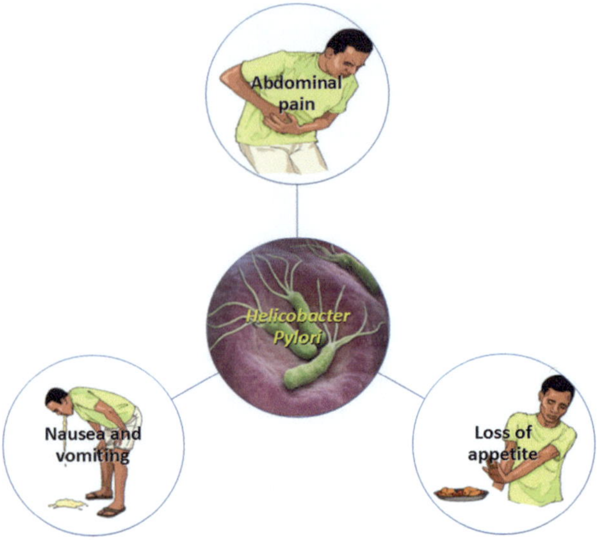

Fig. 5.9 Symptoms of *H. Pylori* infection

tract is associated with symptoms like stomach pain, loss of appetite, nausea, vomiting, and belching. *H. pylori* is the leading cause of global peptic ulcer and gastritis, presenting as gnawing or burning epigastric pain. Infected people have a two- to sixfold higher possibility of developing gastric cancer and lymphoma of the type of mucosal associated lymphoma compared to healthy people.

5.6.12 Polio Virus

Poliomyelitis is caused by polio virus and is considered as an infectious disease that is crippling and potentially deadly. There are three wild types of polio virus (WPV)—type 1, type 2, and type 3. It spreads across people and can penetrate the brain and spinal cord of an infected person, causing paralysis. Polio virus infections are mostly eradicated from the developed and developing countries [20].

5.6.13 Coxsackie Virus A

Coxsackie virus A is the causative agent for hand, foot, and mouth disease (Fig. 5.10) characterized by small red spots developing to blisters on palm, soles, and oral cavity [21]. Coxsackie viruses are among the leading causes of aseptic meningitis. Symptoms include muscle aches, headache, high fever, and nausea and some also develop a sore throat and abdominal discomfort in severe cases of infection.

5.6.14 Coxsackie Virus B

Coxsackie B is a group of six serotypes of Coxsackie virus, a pathogenic enterovirus, that triggers illness ranging from gastrointestinal distress to full-fledged

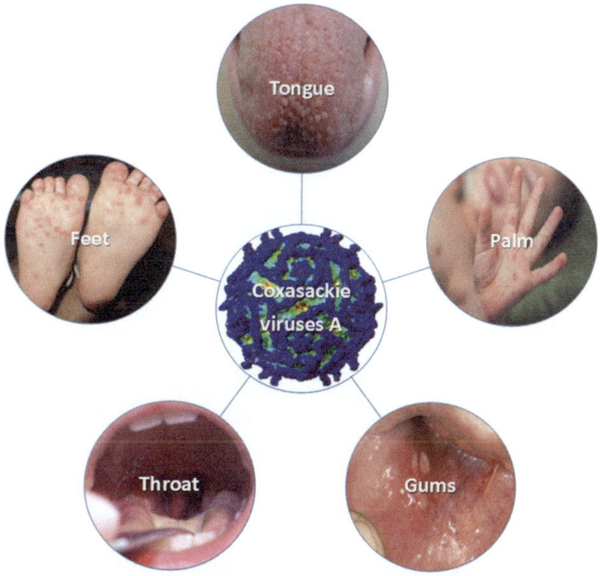

Fig. 5.10 Human infection of Coxasackie viruses A causing Hand Foot and Mouth disease

pericarditis and myocarditis. Coxsackie virus B is the causative agent for pleurodynia and is characterized by intense chest pain and pleural inflammation. Common symptoms include include fever, headache, sore throat, gastrointestinal distress, extreme fatigue, as well as chest and muscle pain.

5.6.15 Echoviruses

The echovirus is one of the many types of viruses that inhabit the gastrointestinal tract. Echoviruses are one of the causative agents for viral meningitis and collectively known as enteroviruses. Enteroviruses are second only to rhinoviruses as the most common viruses in people. Echoviruses are commonly known to cause a range of illnesses starting from febrile infection to serious and potentially fatal conditions including aseptic meningitis, encephalitis, paralysis, and myocarditis.

5.6.16 Rotaviruses

Rotavirus is a contagious gastroenteritis virus. Rotavirus causes a contagious infection which leads to diarrhea (Fig. 5.11) in infants and children around the globe [22]. Severe watery diarrhea, vomiting, fever, and abdominal pain are all symptoms. The most likely diseases of rotavirus are infants and young children. It can get seriously dehydrated and must be admitted to hospital and even die.

5.6.17 Adenoviruses

Adenoviruses cause bronchitis, pneumonia, diarrhea, fever, gastroenteritis, and neurological diseases. They are primarily spread by infected patient's stool contaminated with water.

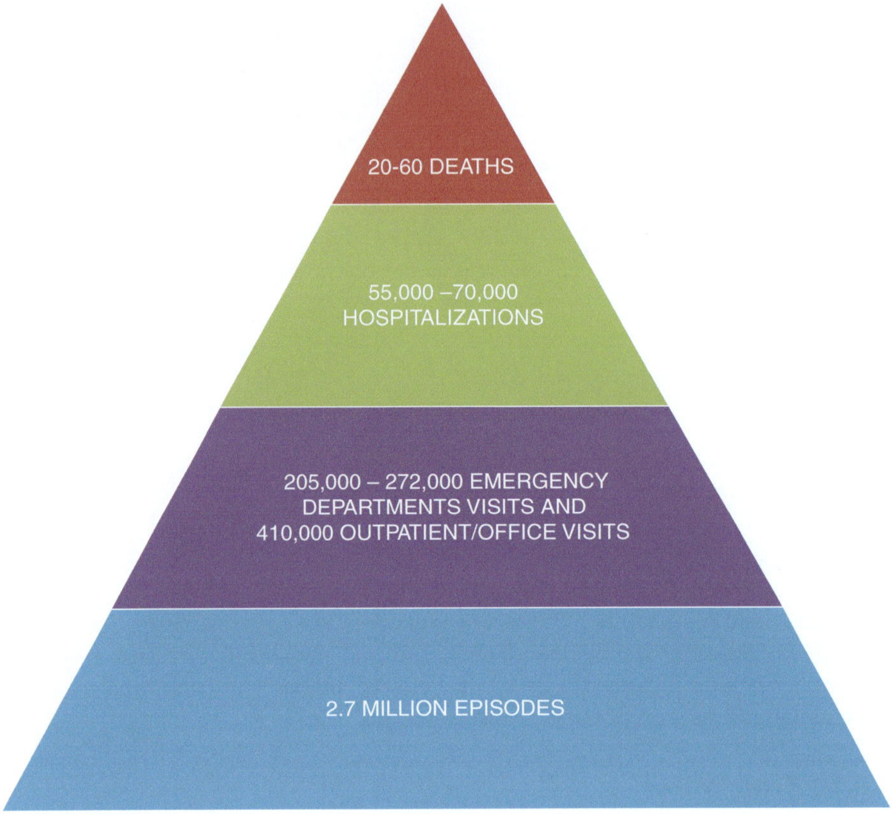

Fig. 5.11 Estimated number of deaths, hospitalizations, emergency department visits and episodes of Rotavirus gastroenteritis among children aged <5 years before introduction of vaccine in United States

5.6.18 Hepatitis A Virus

Hepatitis A virus causes liver disease known as hepatitis A, which is characterized by vomiting, dark urine, and jaundice. The virus is mainly transmitted through contaminated water or food and feces of an infected person (Fig. 5.12). The disease is intimately linked to insecure water or food and poor hygiene. Hepatitis A is an autonomous condition, which does not lead to chronic infection.

5.6.19 Hepatitis E Virus

Hepatitis E virus causes liver disease which is known as hepatitis E. Viral hepatitis occurs mostly in young and middle-aged people and a high mortality rate is significantly seen in pregnant women. Hepatitis E is primarily transmitted through contaminated water via the fecal-oral route. Hepatitis E is found globally but in East and South Asia the prevalence is highest.

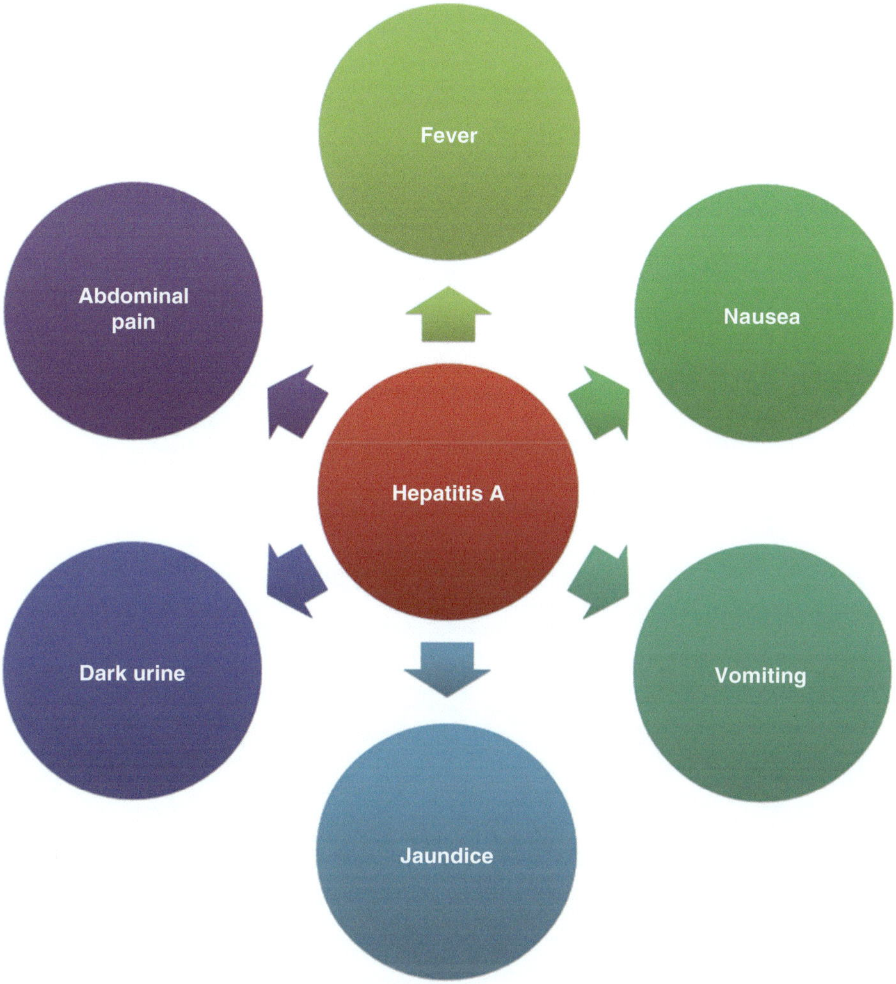

Fig. 5.12 Symptoms during Hepatitis A virus infection

5.6.20 Norovirus

Norovirus is contagious and also known as winter vomiting bug, characterized by diarrhea and inflammation in inner lining of stomach [23]. Globally, about one out of every five cases of acute gastroenteritis that leads to diarrhea and vomiting is caused by norovirus. Norovirus can be transmitted through contaminated water or food and direct contact with previously infected individuals.

5.6.21 *Cryptosporidium hominis*

There are many *Cryptosporidium* species which infects humans and a wide variety of animals. Cryptosporidiosis caused by *Cryptosporidium hominis* is

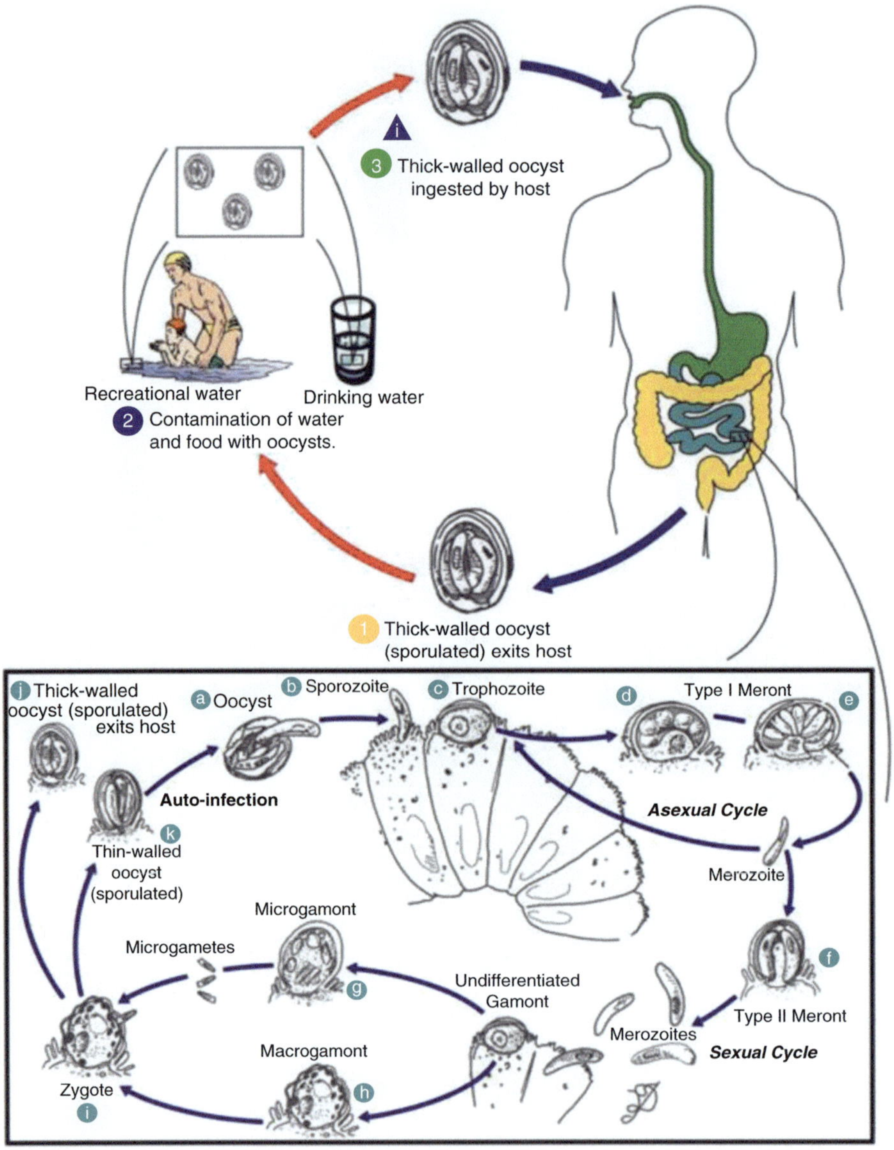

Fig. 5.13 Cryptosporidium lifecycle [Credit: CDC]

characterized by diarrhea (Fig. 5.13), cramps in stomach, fever, weight loss, vomiting, nausea, and loose motions [24]. The parasite is sheltered from the outside of the body, enabling it to survive for long periods and can tolerate chlorine disinfection.

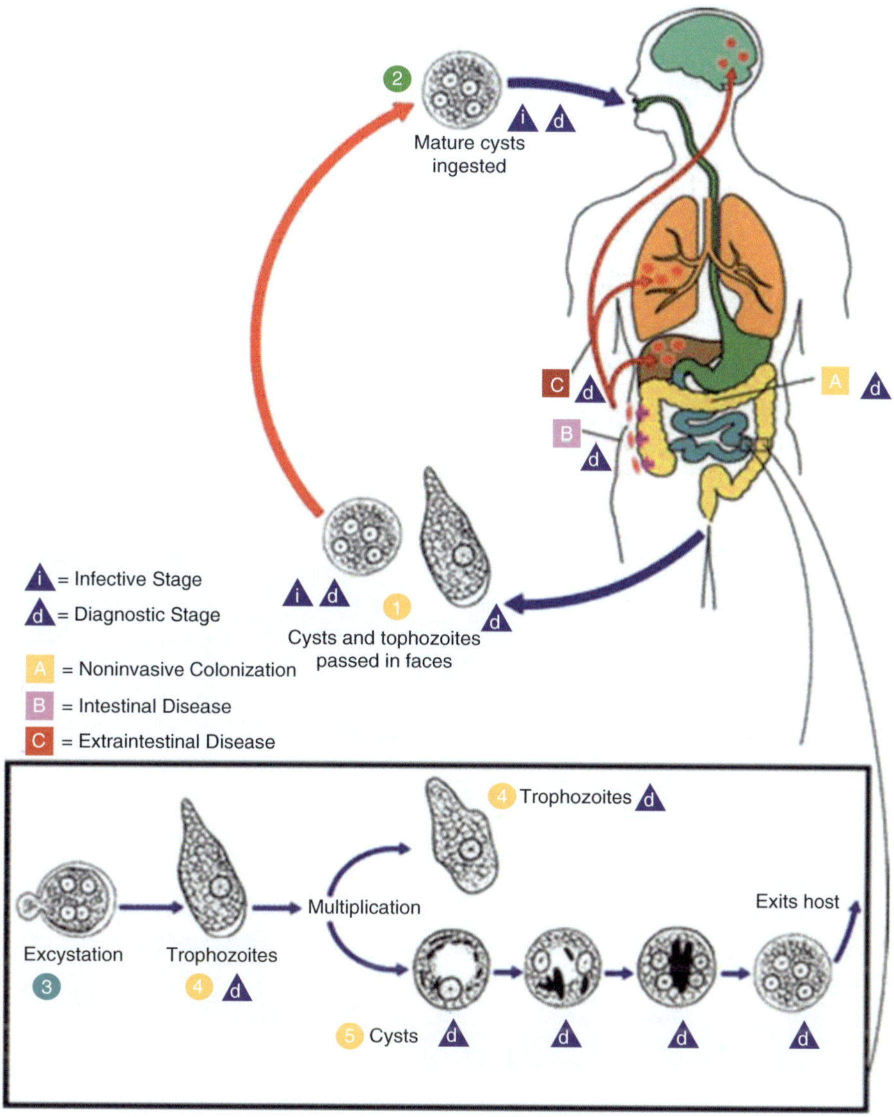

Fig. 5.14 Life cycle of *Entamoeba histolytica* [Credit: CDC]

5.6.22 *Entamoeba histolytica*

Entamoeba histolytica is responsible for amoebiasis or gastrointestinal infection (Fig. 5.14), characterized by diarrhea or dysentery with blood [25]. It can transmit through the fecal-oral route, either directly through person-to-person contact or indirectly through drinking water or eating food contaminated with fecal matter. Acute amoebiasis may occur with frequent, small, and often bloody stools as diarrhea or dysentery. Gastrointestinal symptoms like fatigue, weight loss, and occasional fever may present in chronic amoebiasis. Amoebiasis may occur extraintestinally when the parasite spreads over other organs, usually liver, causing amebic liver abscess. Fever and upper right quadrant abdominal pain are present in the amoebic liver abscess.

5.6.23 *Giardia lamblia*

Giardia lamblia is the causative agent of a parasitic infection giardiasis (Fig. 5.15), characterized by anorexia, weakness, diarrhea, bloating, abdominal cramps, and weight loss [26]. Infection is usually ingested by *G. intestinalis* cysts in water or

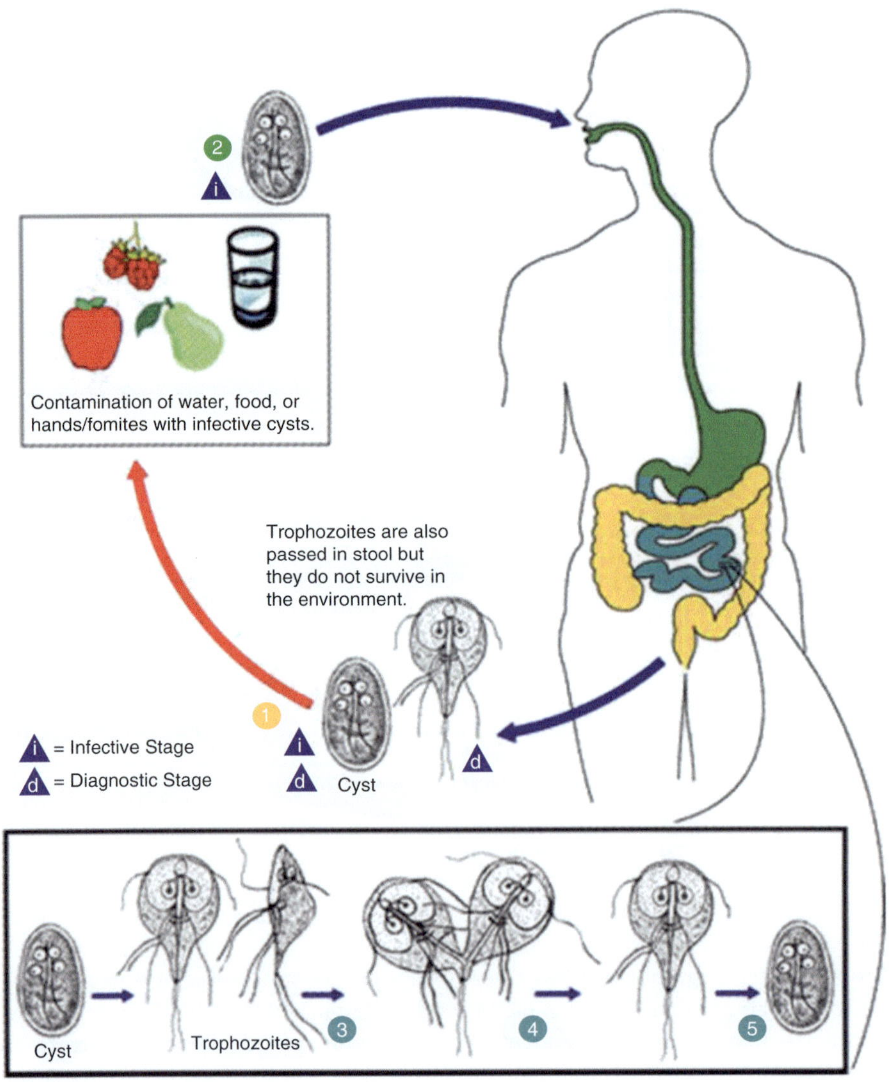

Fig. 5.15 Life-cycle of *Giardia lamblia* [Credit: CDC]

food contaminated with fecal matter of infected humans or animals (including both non-filtered drinking water and recreational waters). Giardia is protected by the external shell, which makes it tolerant to chlorine disinfection and allows it to survive outside the body for a long time.

5.6.24 *Naegleria fowleri*

Naegleria fowleri is the etiological agent of a brain infection primarily brain meningoencephalitis which is characterized by frontal headache, stiff neck (Fig. 5.16), nausea, fever, seizures, vomiting, and hallucinations [27]. When contaminated

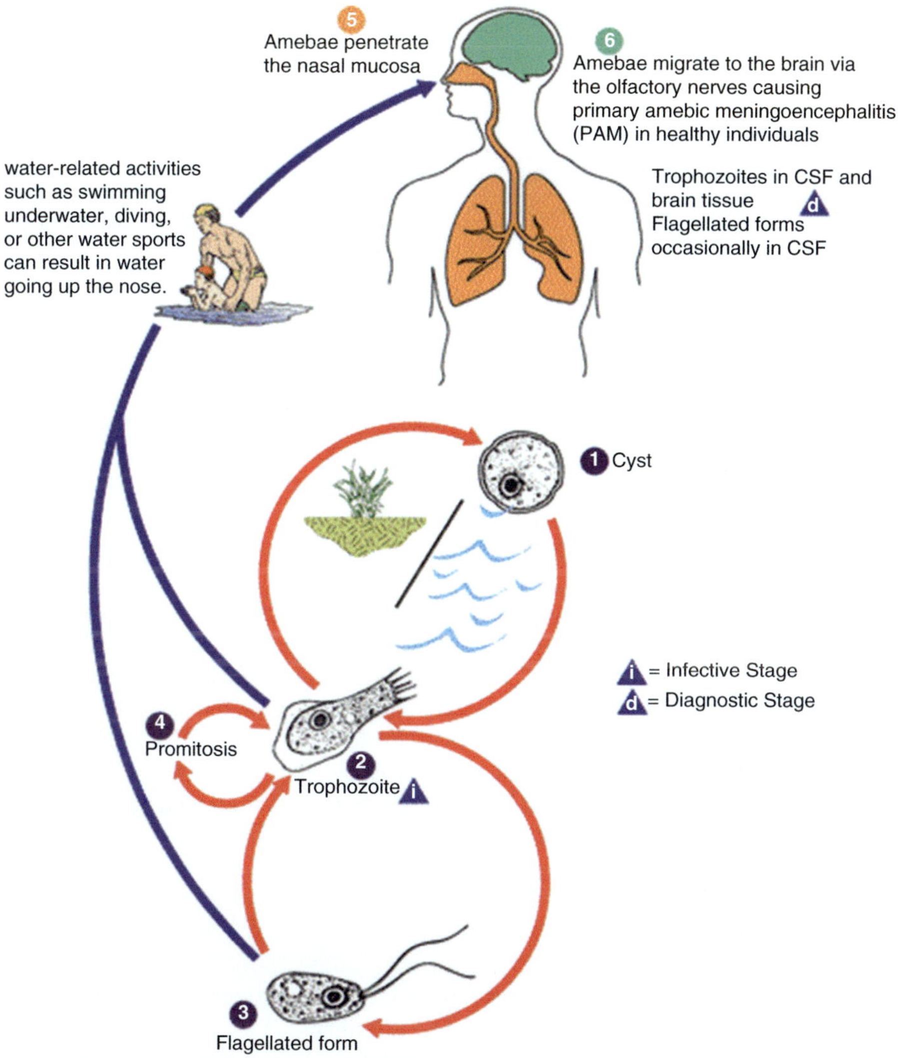

Fig. 5.16 The life cycle of *Naegleria fowleri* human infection [Credit: CDC]

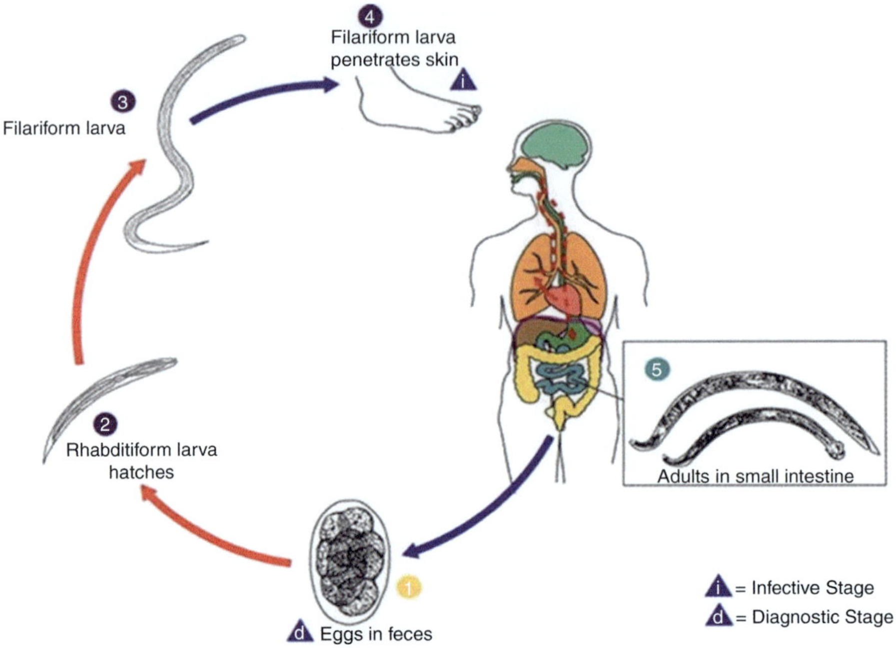

Fig. 5.17 Human infection of *Ascaris lumbricoides* [Credit: CDC]

water enters through the nose into your body, it usually infects humans. Upon entering of the amoeba via nasal route it goes to the brain where primary amebic meningoencephalitis (PAM) occurs which is usually fatal. Infection usually happens if people go to warm, cool places like lakes or rivers to swim or dive.

5.6.25 *Ascaris lumbricoides*

Ascaris lumbricoides or roundworms cause itching, irritability, nausea, vomiting, diarrhea, and difficulty in breathing [28] (Fig. 5.17). Ascaris are parasites and eggs are passed through feces of infected persons. Ascaris are common in the intestine. Eggs are deposited on the soil if an infected person defecates outside or if he or she uses feces as a fertilizer. The eggs can then develop into an infectious form of the parasite. The intake of eggs causes ascariasis. This can happen if you put your hands or fingers on your mouth, or if you consume vegetables or fruits that have not been carefully cooked, washed, or peeled.

5.6.26 Malaria

The bite of infected female Anopheles mosquitoes is the rationale of malaria transmission. Four types of malaria parasites infect humans: *P. falciparum, P. vivax, P. malariae*, and *P. ovale*. People infected with malaria parasites typically become sick

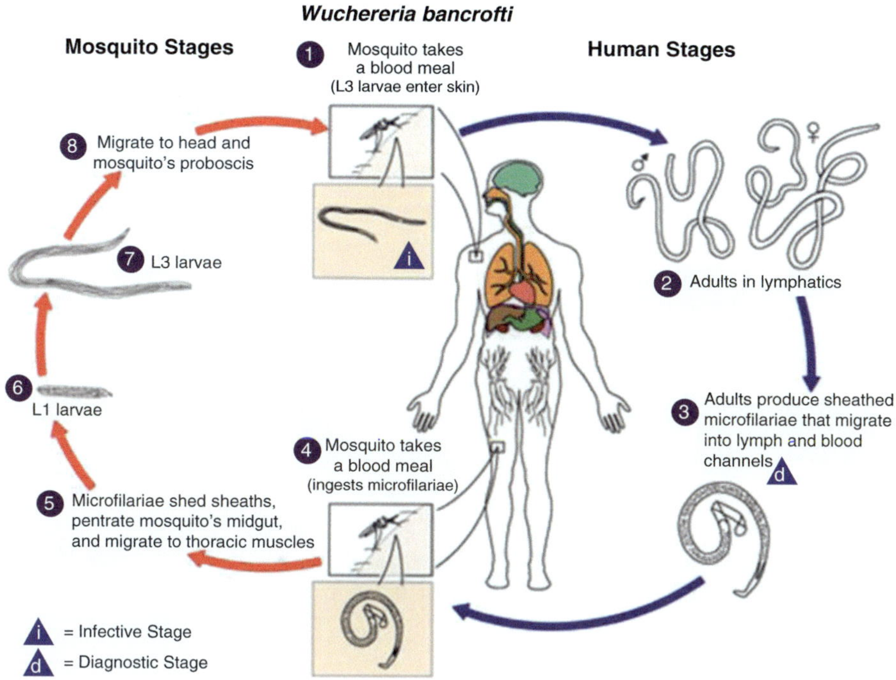

Fig. 5.18 Life cycle of Wuchereria bancrofti [Credit: CDC]

with chills, high-grade fever, and flu-like symptoms and may lead to death in severe conditions like cerebral malaria.

5.6.27 Filaria

Filariasis is a neglected tropical disease caused by parasite *Wuchereria bancrofti*, transmitted by mosquito. Filarial infection can result in a range of symptoms like genital diseases, lymphedema, and elephantiasis (Fig. 5.18). Lymphatic filariasis is the worldwide leading cause of permanent disability. Public health actions have proven to be important for reducing new cases where aerosols, mosquito repellents, creams, spools, and nets have been used and mosquito breeding has been prevented with better hygiene and sanitation practices [29].

5.6.28 Japanese Encephalitis

Japanese encephalitis (JE), a mosquito-borne flavivirus, is the major cause of viral encephalitis in India and Southeast Asia. JEV spreads in humans through infected mosquito bites primarily by *Culex tritaeniorhynchus* (Fig. 5.19). Predominantly human infection is nonsymptomatic or may lead to mild symptoms. Some of the infected individuals develop severe illness with a range of symptoms such as fever,

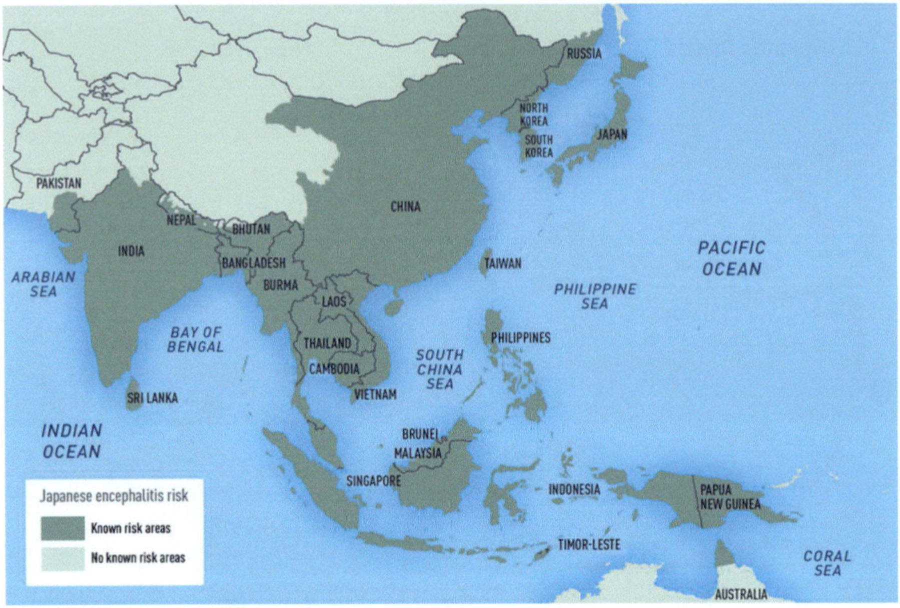

Fig. 5.19 Risk areas for Japanese encephalitis [Credit: CDC]

headache, and vomiting leading to convulsions and altered sensorium. According to the WHO there is no specific treatment available for JE [30]. However, several molecules have been shown to be effective against JEV but none of them are FDA approved or available for the patient use as in dose-dependent manner. The failure to develop effective antiviral drugs against JEV involves several factors including limited in-vitro and in-vivo models such as cell culture, animal models etc. In addition, issues associated with challenging drug trails designs and severe conditions of JEV infected patients including etiological heterogeneity of encephalitis are the reasons for unavailability of effective drugs.

5.6.29 Dengue Fever

Dengue is a flavivirus infection, mainly transmitted by bite of Aedes mosquitoes and responsible for 50–100 million cases annually in over 100 endemic countries. *Aedes aegypti* is the predominant mosquito vector for the viral transmission. The infected individual represents with a range of signs and symptoms like high fever, headache, body ache, severe joint pain, rashes, severe eye pain, weakness, loss of appetite, nausea, and vomiting (Fig. 5.20). Severe case of dengue is defined as dengue hemorrhagic fever. No special treatment is available and the majority of cases require hospitalization, where fluid replacement therapy is prescribed with the use of analgesics (pain relievers) like acetaminophen.

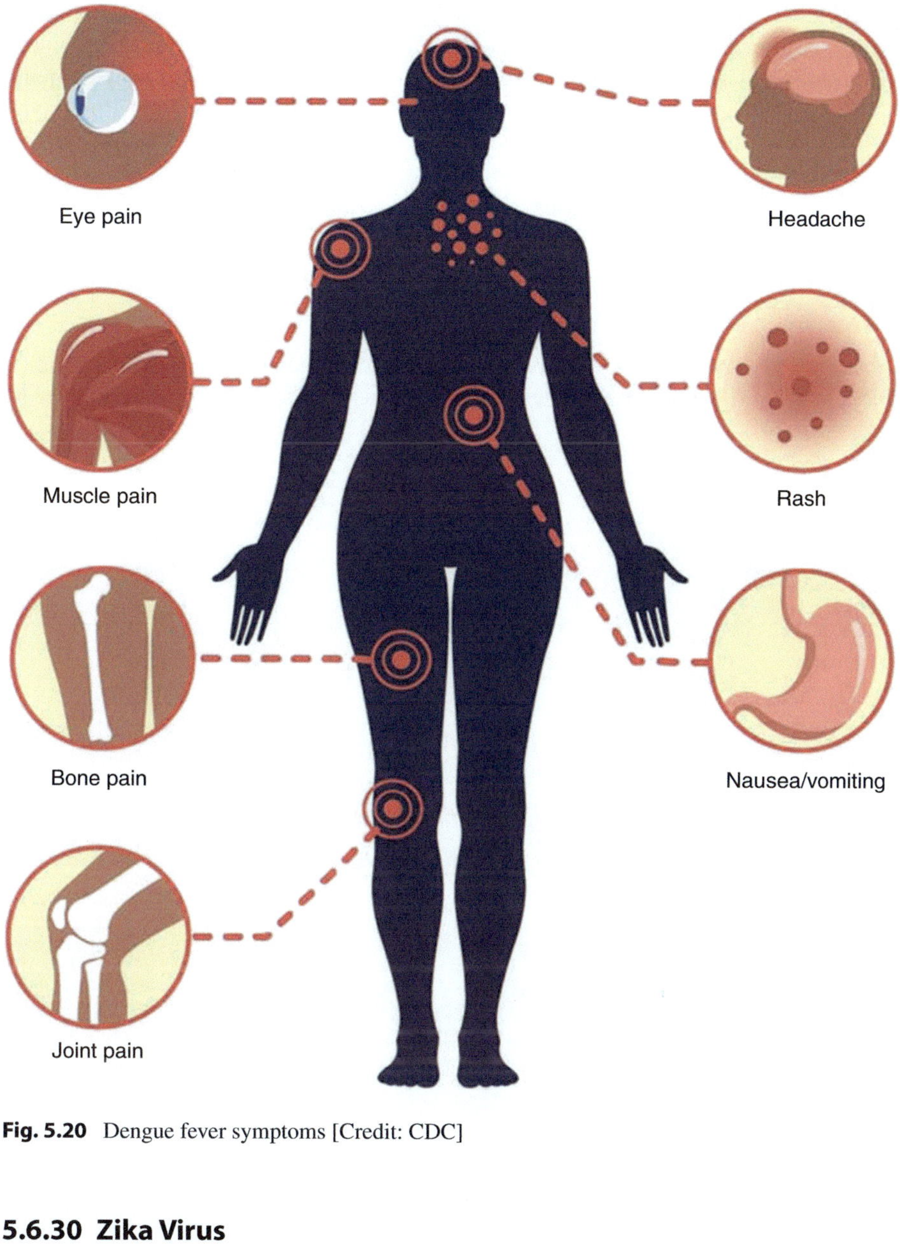

Fig. 5.20 Dengue fever symptoms [Credit: CDC]

5.6.30 Zika Virus

Zika is a mosquito-borne flavivirus infection which is primarily transmitted by infected mosquito bites of species *Aedes aegypti* and *Aedes albopictus*. Mosquito-independent transmission of Zika virus has been identified where sexual transmission and transplacental transmission have been shown to be distinctive among flaviviruses. The symptoms represented by infected individuals are fever, skin rashes, conjunctivitis, muscle pain, and headache which are very similar to dengue symptoms [31]. Infected pregnant women may give rise to child birth with congenital brain abnormalities including microcephaly (Fig. 5.21) and Guillain-Barre syndrome. According to WHO, no specific treatment is available for Zika virus. Recently three cases have been reported from India suggesting that there is a need for urgency in terms of prevention and control of Zika virus infection.

5.6.31 West Nile

West Nile virus (WNV) is a flavivirus transmitted by tick or mosquito (Fig. 5.22). WNV has been shown in a limited number of cases to spread from the mother to

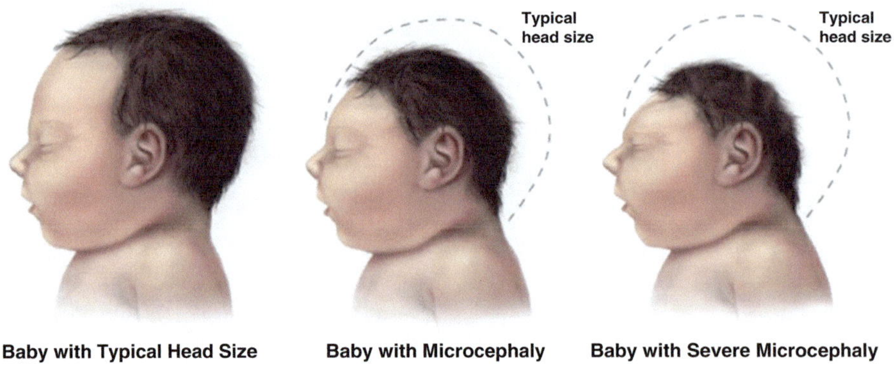

Fig. 5.21 Microcephaly caused by Zika virus infection [Credit: CDC]

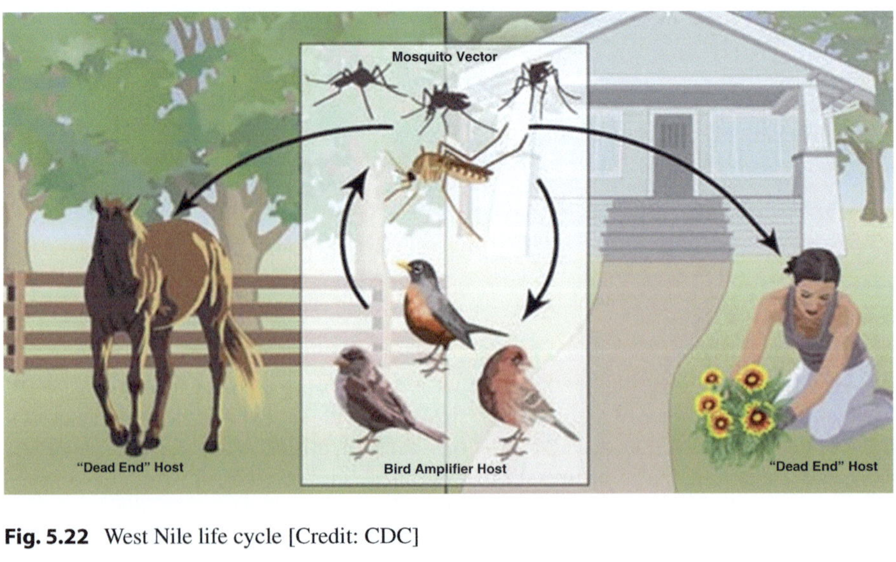

Fig. 5.22 West Nile life cycle [Credit: CDC]

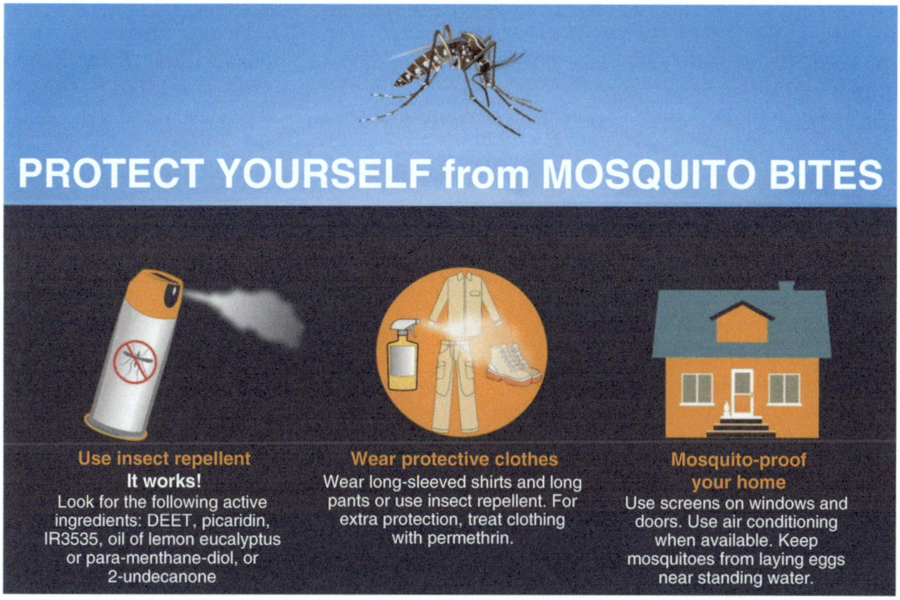

Fig. 5.23 Preventive measures of yellow fever

unborn child and in infants via lactation. WNV may also be transmitted through blood transfusions and organ transplantation. West Nile virus infection may lead to various symptoms like high fever, disorientation or confusion, severe headache, stiff neck, stupor or coma, partial paralysis or muscle weakness, tremors, or muscle jerking and seizures [32]. No vaccine or antivirals are available for the treatment of WNV infection [33].

5.6.32 Yellow Fever

Yellow fever is an acute viral hemorrhagic disease spread by infected mosquitoes [34]. The "yellow" in the name refers to the jaundice that affects some patients (Fig. 5.23). The lack of specific antiviral for the treatment of yellow fever instigates towards adapting other traditional medicines [33].

References

1. Morse SS (1995) Factors in the emergence of infectious diseases. Emerg Infect Dis 1(1):7–15
2. Chang KP (2014) Vaccination for disease prevention and control: the necessity of renewed emphasis and new approaches. J Immunol Immunotech 1(1). https://doi.org/10.17653/2374-9105.SSe0001
3. Lindahl JF, Grace D (2015) The consequences of human actions on risks for infectious diseases: a review. Infect Ecol Epidemiol 5:30048

4. Theron J, Cloete TE (2002) Emerging waterborne infections: contributing factors, agents, and detection tools. Crit Rev Microbiol 28(1):1–26
5. Cohen ML (1998) Resurgent and emergent disease in a changing world. Br Med Bull 54(3):523–532
6. Woolhouse ME, Gowtage-Sequeria S (2005) Host range and emerging and reemerging pathogens. Emerg Infect Dis 11(12):1842–1847
7. Parry CM, Hien TT, Dougan G, White NJ, Farrar JJ (2002) Typhoid fever. N Engl J Med 347(22):1770–1782
8. Rowe B, Ward LR, Threlfall EJ (1997) Multidrug-resistant Salmonella typhi: a worldwide epidemic. Clin Infect Dis 24(Suppl 1):106–109
9. Thielman NM, Guerrant RL (2004) Clinical practice. Acute infectious diarrhea. N Engl J Med 350(1):38–47
10. Clemens JD, Nair GB, Ahmed T, Qadri F, Holmgren J (2017) Cholera. Lancet 390(10101):1539–1549
11. Waldman RJ, Mintz ED, Papowitz HE (2013) The cure for cholera—improving access to safe water and sanitation. N Engl J Med 368(7):592–594
12. Ochoa TJ, Contreras CA (2011) Enteropathogenic *Escherichia coli* infection in children. Curr Opin Infect Dis 24(5):478–483
13. Cover TL, Aber RC (1989) Yersinia enterocolitica. N Engl J Med 321(1):16–24
14. Ford AC, Lacy BE, Talley NJ (2017) Irritable bowel syndrome. N Engl J Med 376(26):2566–2578
15. Chuang CH, Wang YH, Chang HJ, Chen HL, Huang YC, Lin TY, Ozer EA, Allen JP, Hauser AR, Chiu CH (2014) Shanghai fever: a distinct Pseudomonas aeruginosa enteric disease. Gut 63(5):736–743
16. Wilcox MH, Cook AM, Eley A, Spencer RC (1992) Aeromonas spp. as a potential cause of diarrhoea in children. J Clin Pathol 45(11):959–963
17. Stout JE, Yu VL (1997) Legionellosis. N Engl J Med 337(10):682–687
18. Haake DA, Levett PN (2015) Leptospirosis in humans. Curr Top Microbiol Immunol 387:65–97
19. McColl KE (2010) Clinical practice. Helicobacter pylori infection. N Engl J Med 362(17):1597–1604
20. Parker EP, Grassly NC (2016) Polio vaccination: preparing for a change of routine. Lancet 388(10040):107–108
21. Harris PNA, Wang AD, Yin M, Lee CK, Archuleta S (2014) Atypical hand, foot, and mouth disease: eczema coxsackium can also occur in adults. Lancet Infect Dis 14(11):1043
22. Parashar UD, Nelson EA, Kang G (2013) Diagnosis, management, and prevention of rotavirus gastroenteritis in children. BMJ 347:7204
23. Lancet Editorial paper (2013) The inexorable progress of norovirus. Lancet Infect Dis 13(2):97
24. Chen XM, Keithly JS, Paya CV, LaRusso NF (2002) Cryptosporidiosis. N Engl J Med 346(22):1723–1731
25. Stanley SL Jr (2003) Amoebiasis. Lancet 361(9362):1025–1034
26. Minetti C, Chalmers RM, Beeching NJ, Probert C, Lamden K (2016) Giardiasis. BMJ 355:5369
27. Centers for Disease Control and Prevention (2017) National Center for Emerging and Zoonotic Infectious Diseases (NCEZID). Division of Global Migration and Quarantine (DGMQ). Parasites—Naegleria fowleri—Primary Amebic Meningoencephalitis (PAM)—Amebic Encephalitis. CDC, Atlanta, GA. https://www.cdc.gov/parasites/naegleria/index.html. Accessed on 11-3-2018
28. Bethony J, Brooker S, Albonico M, Geiger SM, Loukas A, Diemert D, Hotez PJ (2006) Soil-transmitted helminth infections: ascariasis, trichuriasis, and hookworm. Lancet 367(9521):1521–1532
29. Ottesen EA, Duke BO, Karam M, Behbehani K (1997) Strategies and tools for the control/elimination of lymphatic filariasis. Bull World Health Organ 75(6):491–503
30. World Health Organization (2015) Japanese encephalitis. Fact sheet no. 386. WHO, Geneva. http://www.who.int/mediacentre/factsheets/fs386/en/. Accessed on 15-3-2018

31. Petersen LR, Jamieson DJ, Powers AM, Honein MA (2016) Zika virus. N Engl J Med 374(16):1552–1563
32. Petersen LR, Roehrig JT, Hughes JM (2002) West Nile virus encephalitis. N Engl J Med 347(16):1225–1226
33. World Health Organization (2011) West Nile virus. Fact sheet no. 354. WHO, Geneva. http://www.who.int/mediacentre/factsheets/fs354/en/. Accessed on 15-3-2018
34. World Health Organization (2018) Yellow fever. Fact sheet. WHO, Geneva. http://www.who.int/mediacentre/factsheets/fs100/en/. Accessed on 15-3-2018

Impact of Climate Change on Water-Associated Infectious Diseases

6

Amrita Haikerwal* and Shailendra K. Saxena*

Abstract

The relation between climate change and water-associated infectious diseases has become much more visible in recent times. Climatic changes are dealt with complex interactions among interconnected components of hydrological cycle such as atmosphere, biosphere, and geosphere. Extreme weather conditions such as storms may pollute the recreational coastal waters enhancing the risk of gastroenteritis and other water-associated infectious diseases. The contamination of drinking water is a prominent source of spreading water-associated infectious diseases during extreme water-related weather events. The water treatment plant system may get compromised due to heavy rainfalls insinuating infiltration of oocysts of cryptosporidium in drinking water reservoirs and remain persistent in water distribution system. Weather conditions influence the life cycle of arthropod vectors by affecting the survival rate, reproduction rates, habitat, abundance, and distribution; likewise it also affects the survival rate and proliferation rate of pathogen within the vector.

Keywords

Climatic changes · Droughts · Increased rainfall · Storms · Floods · High temperature and humidity

*Contributed equally as first author

A. Haikerwal · S. K. Saxena (✉)
Centre for Advanced Research, Faculty of Medicine, King George's Medical University, Lucknow, India
e-mail: shailen@kgmcindia.edu

S. K. Saxena (ed.), *Water-Associated Infectious Diseases*,
https://doi.org/10.1007/978-981-13-9197-2_6

6.1 Introduction

Since nineteenth century, humans were aware that climatic conditions cause spread of epidemic diseases, even when the significance of infectious pathogens was unknown. An anthropogenic activity such as fossil fuel ignition that produces greenhouse gases over a long period has resulted in global change in characteristic weather. The effects of change in characteristic weather have led to decline in ecosystems, altered weather conditions, depletion of stratospheric ozone layer, and decrease in biodiversity to an unprecedented extent. The global mean surface temperature has increased by 0.74 °C in the past century, whereas the worldwide sea level has increased by 1.8 mm/year since 1961 and the Arctic sea ice meltdown is decreasing by 2.7% per decade [1]. Other environmental changes include shrinking of mountain glaciers, altered rainfall patterns, and higher frequency of extreme weather conditions originated by natural disasters like hurricane, flood, tsunami, and cyclone, detrimentally affecting human health as well as flora and fauna.

The relationship between climate change and infectious diseases particularly water-associated infectious diseases has recently become much more evident. It influences the proliferation rate and reproduction rates of pathogens at higher temperatures; enhances the range of vector-borne infectious diseases during heavy precipitation; extends breeding period and distribution of disease, migration of insect vectors, and host animal reservoirs; and increases the burden of disease by making humans as dead-end hosts [2]. Thus, dependence of climate change and infectious diseases requires regular assessment and analysis of incidence rate of the disease, prevalence of pathogens and their genotypes in particular region, distribution of outbreaks, and potential tropical vector species. The integration of environmental and epidemiological networks will strengthen the knowledge about the connection of climate change and water-associated infectious diseases that will help in taking well-informed decisions for public health.

6.2 Climate Change and Water-Associated Infectious Diseases

Climatic changes are dealt with complex interactions among interconnected components of hydrological cycle such as atmosphere, biosphere, and geosphere. The interference with these factors alters the average meteorological measures which influence the prevalence of catastrophic events such as increased rainfall, storms, droughts, and floods (Fig. 6.1). Such disasters hamper the quality, accessibility, and availability of water imposing a threat to humankind. Water-associated infectious diseases in humans mostly occur by exposure of impure drinking water. Vector-borne diseases are another aspect of water-associated infectious diseases where excessive stagnant water inculcates germination of mosquitoes that might be responsible for spreading of pathogens causing dengue, malaria, chikungunya, Japanese encephalitis, West Nile fever, Zika, etc.

Fig. 6.1 Climate change and associated WAID

6.3 Recreational Usage of Water

With the rising mean surface temperatures, rapid evaporation of water occurs which later condenses and triggers heavy intense precipitation [3]. Consequently, insinuating runoff, inundations and deposition of nutrients, toxic chemicals, and microbial growth in coastal waters may adversely harm marine life and public health. Usage of coastal watersheds for recreational activities had enhanced development and urbanization of coastal areas. Extreme weather conditions such as storms may pollute the recreational coastal waters enhancing the risk of gastroenteritis and other water-associated infectious diseases. During the El Nino winters in southern California, the precipitation rate is high compared to the La Nina winters instigating higher risk of symptoms associated with water-associated infectious diseases [4]. As the excessive rain is adverse so are the long periods of hot weather conditions; this provides the favorable temperature for the germination and growth of microorganisms and algal species in water bodies. For instance, in hot summers of 2006, a case study was reported of the Baltic and the North Sea where bacterial species such as *Vibrio cholerae* and *Vibrio vulnificus* are indigenous and showed increased incidences of growth of these bacteria causing open wounds that necrotize and cause severe sepsis [5, 6]. *Cryptosporidium* is responsible for 52% of outbreaks in the recreational context, as stated in the Morbidity and Mortality Weekly Report (MMWR). *Cryptosporidium* is a parasite which can survive in the chlorinated water (chlorine-tolerant parasite) for more than 10 days. Similarly, the main cause of parasite infection in the USA is *Giardia intestinalis* estimated at 1.2 million cases and 3581 reported hospitalizations annually [7]. Several routes of giardiasis infection have been documented but the most effective is water-borne (74.6%) which is related with pure water and recreational sources primarily swimming pools [8].

6.4 Drinking Water

Water plays a crucial role in the transmission of plethora of infectious diseases either directly or indirectly, enhancing the global burden of diseases. A global electronic system, ProMED, from 1910 to 2010 has reported 235 water-associated outbreaks including extreme water-related weather events where elevated precipitations and flood were most commonly reported [9]. In these water-borne outbreaks, majority of pathogens belong to *Vibrio* spp. and *Leptospira* spp. The contamination of drinking water was a prominent source of spreading water-associated infectious diseases during extreme water-related weather events. The water treatment plant system may get compromised due to heavy rainfalls insinuating infiltration of oocysts of cryptosporidium in drinking water reservoirs and remain persistent in water distribution system [10, 11]. During low rainfalls or extended drought conditions, as the volume of water decreases in river, it becomes concentrated with pathogens that might be difficult to remove by the water treatment plants to make it fit for drinking [12, 13]. In developing countries where 90% of deaths are from children, according to the World Health Organization (WHO), 99.8% is due to poor water, sanitation, and hygiene conditions [14]; In addition, it was reported that 3.6% of DALY estimated global disease burden of the total adjusted disability year (TDAY) is due to 1.5 million deaths every year due to diarrheal diseases. The dominant reason behind such high toll is unsafe drinking water especially in low-income countries and being incompetent in practicing good standards in water, sanitation, and hygiene [15]. Numerous water-associated infectious diseases caused by contamination of drinking water with animal or human excreta are typhoid fever (*Salmonella typhi*), giardiasis (*Giardia lamblia*), cholera (*Vibrio cholerae*), poliomyelitis (polio viruses), gastroenteritis (rotaviruses), and amoebic dysentery (*Entamoeba histolytica*) [16].

6.5 Mosquito-Borne Diseases

Weather conditions influence the life cycle of arthropod vectors by affecting the survival rate, reproduction rates, habitat, abundance, and distribution; likewise it also affects the survival rate and proliferation rate of pathogen within the vector. The predominant vectors are mosquitoes which can transmit the etiological causative agents to the humans upon biting. High temperature and humidity are correlated with high reproduction rate of various spp. of mosquitoes. During the epidemic period, it has been observed that rainfall is low and mosquitoes such as *Culex pipiens* concentrate around the water sources with their avian hosts leading to the spread of arbovirus multiplication. Numerous arboviral diseases caused by flavivirus belong to the family Flaviviridae like Japanese encephalitis, Zika, dengue, West Nile, and yellow fever; apart from these other diseases transmitted by mosquitoes are malaria, chikungunya fever, filariasis, etc. West Nile fever is caused by the West Nile virus, which mainly infects birds and humans through the bite of *Culex pipiens* infected with the virus [17]. Most of the cases are asymptomatic and the etiological data of infected human beings are largely unknown. Several models have linked the meteorological data with entomological data to understand the pattern of West Nile virus outbreak. The burden of dengue is increasing worldwide as the incidence rate of this arboviral disease is rapidly enhancing in human beings. Urbanization, globalization, and easier means of travel and transport have led to the spread of dengue virus throughout the globe. Its principal vector is *Aedes aegypti* commonly found in urban regions; however another competent vector is *Aedes albopictus*. Various epidemiological studies have reported that temperature plays a significant role in transmitting dengue in urban locations [18]. Chikungunya fever is another arboviral disease belonging to the genus Alphavirus and family Togaviridae, spread to human by the bite of mosquitoes infected with the virus [19].

6.6 Strategies to Improve Public Health During Extreme Weather Conditions

Development of sustainable strategies is immediately required to tackle the technical challenges unleashed at the time of erratic and extreme weather conditions to prevent damage to public health. Linking environmental and epidemiological research networks for intensive analysis will speed up the understanding of changes in characteristics of weather relationship with infectious diseases and improve public health [20]. Several steps elucidated by leading public health agencies are enlisted below:

6.6.1 Monitoring and Surveillance

Regular collection, assessment, analysis, and understanding of data associated with change in characteristics of weather are done in order to correlate with the epidemiological data of infectious diseases. Sentinel surveillance system which collects high-quality data regarding the infectious disease with reliability of accuracy is manifested from specific geographical region. Mandatory vector surveillance monitors the distribution of specific species present in a particular area. Prescribed or non-prescribed drugs from pharmacy should be monitored to ascertain the diagnosis of prevalent diseases. There should be instantaneous data collection of patients diagnosed with infectious diseases and reporting of hospital admissions and autopsy reports. Surveillance of dead birds or symptoms indicating infectious disease should be done.

6.6.2 Investigation of Outbreaks

Emerging tropical diseases are inspected regularly and rapid response is shown to control the dissemination of outbreak by following the integrated methods to control vector-borne infectious diseases. Adequate supply of medicinal personnel and medication such as antivirals and vaccines is ensured to effectively control the situation.

6.6.3 Rightly Inform, Educate, and Empower About Public Health Education

Well-organized information regarding safe drinking water, sanitation, and hygiene is urgently needed to be disseminated. Recreational activities should be carried out with clean water. During hot weather conditions, awareness regarding proper handling of food and water is required. Health education and timely dissemination of strategies to control vectors using integrated pest control management, removal of stagnant water, and usage of nets and mosquito repellents should be promoted.

6.6.4 Development of Interagency and Community Partnerships

Connecting people of different sectors and disciplines will instigate unprecedented ideas and approaches that will help in solving a plethora of problems during outbreaks. For instance, civil engineers and sanitation department personnel could solve the problem of degrading drinking water quality; likewise, entomologists working in association with agronomist could promote vector surveillance strategies. Campaign should be initiated to promote vaccination during outbreaks with the media advocates.

6.6.5 Development of Policies and Enforcement of Laws

Infectious diseases sensitive to altering climatic conditions should be put on notifiable disease list. There should be strategies to ensure preparedness and management at regional, national, and international strata. Mandatory enforcement of international health regulations should be focused where a change in climatic conditions that may lead to international crisis has to be reported immediately.

6.6.6 Enhance Accessibility to Care and Competent Personnel

Patients infected with infectious diseases should be provided with immediate treatment to culminate propagation of outbreak like in case of malaria, dengue, and chikungunya. There should be accessibility to vaccination, medication, and care to one and all. Trained and well-equipped healthcare competent workforce is required to be prepared during emergency situations.

6.6.7 Research and Evaluation

Sympathetic relation between climate change and infectious diseases needs to be explored. Parameters of vector ecology affected by climate change are the areas of research. There should be insights into the populations or communities vulnerable to the outbreaks. Portable ready-to-use kits, diagnostic tests, and vaccines should be developed to suffice the needs at the time of emergency. Simulations and models are required to predict the burden of disease with changing climatic parameters.

6.7 Extreme Weather Events and Water-Associated Infectious Diseases

Intense weather events are the meteorological events that affect the local community or an ecosystem. This may include high temperatures, winds, and heavy precipitation turning into natural disasters such as drought and floods. Every year due to such events death toll rises up to thousands worldwide, and these weather events also disrupt local communities, with spread of infectious diseases causing outbreaks that becomes a torment situation. Climate change affects the intensity, timing, duration, and frequency of extreme weather events [21].

6.7.1 Flood

The worldwide recurrence of cases of floods is quite high and is also responsible for the death and damage of millions. With increasing urbanization and to accommodate the population, individuals live near coastal regions and flood-prone areas. This decreases the capacity of land to absorb water at the time of heavy precipitation, which makes them prone to flash floods. Extensive and robust drainage systems should be built to tackle with the problems of runoff and ensure minimum loss of lives from rapidly increasing levels of water during floods. During floods, the immediate concern is to protect survivors and people who are displaced from getting infected with infectious diseases such as measles, diarrhea, malaria, and dengue that might be responsible for high death tolls [22].

6.7.2 Tsunami

Tsunami is a series of ocean waves that develop destructive surges of water which sometimes reach heights of 100 ft or 30.5 m, which occurs due to earthquake under or near the ocean, or due to volcanic eruptions. Consequently, flooding occurs in the nearby regions but is limited to a few 100 m of coastal areas. The risk of infection could be mitigated as destructed area is limited and survivors could be rescued and provided with shelters. In 2004 Indian Ocean tsunami, polymicrobial infections were frequent with marine pathogens such as *Aeromonas* spp. and *Pseudomonas* spp., enteric bacteria found in sewage and wound, and systemic infections were mainly caused by *Plesiomonas* spp., *Burkholderia* spp., and *Staphylococcus aureus* [24].

6.7.3 Hurricane

Hurricanes are destructive storms that develop over warm seawater and move towards the coastal areas. They cause heavy precipitation, coastal region and inland flooding, and strong wind currents, and might lead to tornadoes and landslides. Hurricanes are devastating as they cause massive disruption such as increased risk of infectious diseases such as Zika, dengue, hepatitis A, and leptospirosis; transmission of diseases in areas not prone to them; inadequacy of drinking water and sanitation; and decreased access to food, medication, and shelter. Hurricane Katrina on the US Gulf Coast was established in 2005 to accommodate more than 200,000 people evacuated in 18 states. Health organizations such as CDC assisted local and regional health departments in controlling the situation by initiating regular surveillance of the infectious diseases and outbreaks, and providing medicinal personnel and competent workforce [25].

6.8 Conclusions and Future Perspectives

Climate change, intense weather events, natural disasters, and water-associated infectious diseases are tied in one closed loop that demands an urgent attention to be considered as a global challenge. Initiatives to inform and educate individuals, policy makers, and health officials and to help them understand about the management of extreme weather events and its aftermath that might involve spread of infectious disease outbreaks are pertinent. Strengthening of surveillance systems to accurately understand the extremities of disasters as well as the incidence rate of infectious diseases is obligatory to tackle the challenges imposed. Vigorous research regarding the impact of climate change on transmission and distribution of infectious diseases, vectors, and emerging pathogens is required. Quantification of the causes of high death tolls during extreme weather events is imperative to understand the management of future events. The most probable causes might be urbanization, infectious disease outbreak, migration and globalization that transmit infectious diseases, climate change, and socioeconomic development. Climate changes could increase breeding of vectors, and thus integrated methods to control vector population should be established.

References

1. Semenza JC, Menne B (2009) Climate change and infectious diseases in Europe. Lancet Infect Dis 9(6):365–375
2. Lafferty KD (2009) The ecology of climate change and infectious diseases. Ecology 90:888–900
3. Karl TR, Trenberth KE (2003) Modern global climate change. Science 302:1719–1723
4. Dwight RH, Baker DB, Semenza JC, Olson BH (2004) Health effects associated with recreational water use: urban versus rural California. Am J Public Health 94:565–567
5. Frank C, Littman M, Alpers K, Hallauer J (2006) Vibrio vulnificus wound infections after contact with the Baltic Sea, Germany. Euro Surveill 11:E060817.1
6. Andersson Y, Ekdahl K (2006) Wound infections due to Vibrio cholerae in Sweden after swimming in the Baltic Sea, summer 2006. Euro Surveill 11:E060803.2
7. CDC (2015) Cryptosporidiosis surveillance—United States, 2011–2012. MMWR Suppl 64(3):15–25
8. Strathmann M, Horstkott M, Koch C, Gayer U, Wingender J (2016) The River Ruhr—an urban river under particular interest for recreational use and as a raw water source for drinking water: the collaborative research project "Safe Ruhr"—microbiological aspects. Int J Hyg Environ Health 219(7 Pt B):643–661
9. Cann KF, Thomas DR, Salmon RL, Wyn-Jones AP, Kay D (2013) Extreme water-related weather events and waterborne disease. Epidemiol Infect 141(4):671–686
10. Kistemann T, Classen T, Koch C et al (2002) Microbial load of drinking water reservoir tributaries during extreme rainfall and runoff. Appl Environ Microbiol 68:2188–2197
11. Lake IR, Bentham G, Kovats RS, Nichols GL (2005) Effects of weather and river flow on cryptosporidiosis. J Water Health 3:469–474
12. Wilby R, Hedger M, Orr HG (2005) Climate change impacts and adaptation: a science agenda for the Environment Agency of England and Wales. Weather 60:206–211
13. Senhorst HA, Zwolsman JJ (2005) Climate change and effects on water quality: a first impression. Water Sci Technol 51:53–59
14. Ashbolt NJ (2004) Microbial contamination of drinking water and disease outcomes in developing regions. Toxicology 198(1–3):229–238
15. Sorenson SB, Morssink C, Campos PA (2011) Safe access to safe water in low income countries: water fetching in current times. Soc Sci Med 72(9):1522–1526
16. White GF, Bradley DJ, White AU (2002) Drawers of water: domestic water use in East Africa. 1972. Bull World Health Organ 80(1):63–73
17. Mackenzie JS, Gubler DJ, Petersen LR (2004) Emerging flaviviruses: the spread and resurgence of Japanese encephalitis, West Nile and dengue viruses. Nat Med 10(12 Suppl):S98–S109
18. Hales S, de Wet N, Maindonald J, Woodward A (2002) Potential effect of population and climate changes on global distribution of dengue fever: an empirical model. Lancet 360(9336):830–834
19. Lambrechts L, Scott TW, Gubler DJ (2010) Consequences of the expanding global distribution of Aedes albopictus for dengue virus transmission. PLoS Negl Trop Dis 4(5):e646
20. Frumkin H, Hess J, Luber G, Malilay J, McGeehin M (2008) Climate change: the public health response. Am J Public Health 98:435–445

21. Semenza JC, Giesecke J (2008) Intervening to reduce inequalities in infections in Europe. Am J Public Health 98:787–792
22. Greenough G, McGeehin M, Bernard SM, Trtanj J, Riad J, Engelberg D (2001) The potential impacts of climate variability and change on health impacts of extreme weather events in the United States. Environ Health Perspect 109(Suppl 2):191–198
23. Lignon BL (2006) Infectious diseases that pose specific challenges after natural disasters: a review. Semin Pediatr Infect Dis 17:36–45
24. Garzoni C, Emonet S, Legout L, Benedict R, Hoffmeyer P, Bernard L, Garbino J (2005) Atypical infections in tsunami survivors. Emerg Infect Dis 11(10):1591–1593
25. Centers for Disease Control and Prevention (CDC) (2005) Infectious disease and dermatologic conditions in evacuees and rescue workers after Hurricane Katrina—multiple states, August–September, 2005. MMWR Morb Mortal Wkly Rep 54(38):961–964

Hazards Associated with Contaminated Water

7

Amrita Haikerwal*, Swatantra Kumar,
and Shailendra K. Saxena*

Abstract

The source of contamination of water could be comprised in four categories such as aesthetic, biological, chemical, and dissolved solids. Developing countries are the birthplace of water-associated infectious diseases as the basic necessities of water, hygiene (WASH), and sanitation are not according to the required standards. The changing source of water bodies changes the pathogens related to water-associated infectious diseases. Various other reasons of contamination of water resources that eventually lead to rise in water-associated infectious diseases are traditional burial of dead body releasing decaying microbial agents in the soil that percolate during rainfall and adulterate the groundwater.

Keywords

Fertilizers · Pesticides · Chlorine · Dissolved solids · Biological contamination WASH

7.1 Introduction

The international sharing of water-related diseases has been classified as water-borne, water-dispersed, water-washed, and water-related based on socio-environmental factors [1]. The growing population density is the culprit behind the rising number of epidemics of water-associated infectious diseases [2]. The source of contamination of water could be comprised in four categories such as aesthetic,

*Contributed equally as first author

A. Haikerwal · S. Kumar · S. K. Saxena (✉)
Centre for Advanced Research, Faculty of Medicine, King George's Medical University, Lucknow, India
e-mail: shailen@kgmcindia.edu

S. K. Saxena (ed.), *Water-Associated Infectious Diseases*,
https://doi.org/10.1007/978-981-13-9197-2_7

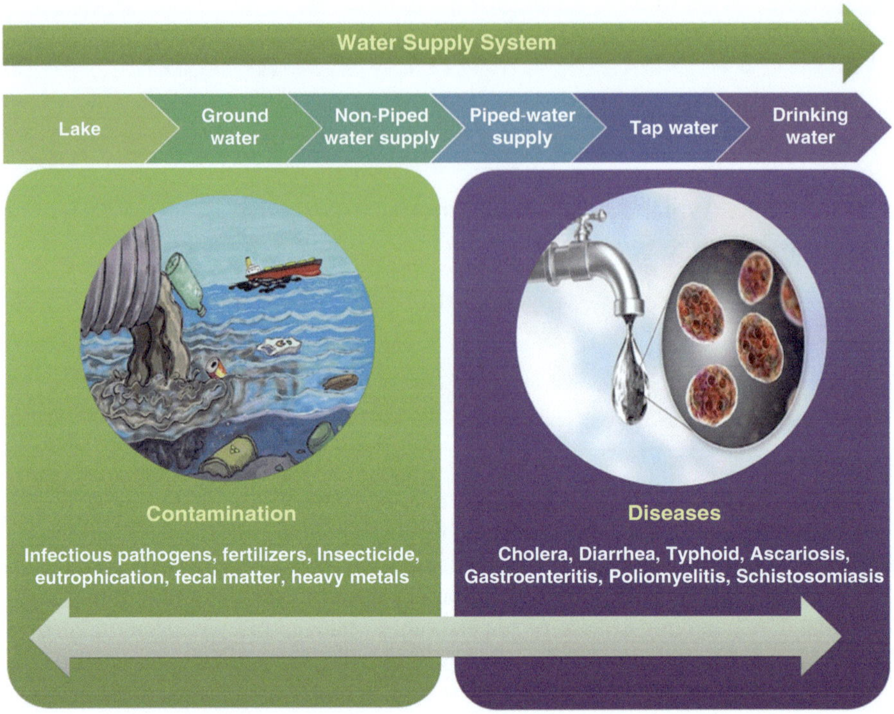

Fig. 7.1 Water supply system from resource to tap. Various modes of contamination of water resources which are carried on by non-piped (wells) or piped water supply system used for drinking might have pathogens capable of causing diseases

biological, chemical, and dissolved solids. Aesthetic contamination of water includes deposition of dirt, minerals, and sediments that brings foul odor in water; biological contamination of water involves pathogens (Fig. 7.1) such as virus, parasites, and bacteria affecting human health; eutrophication by cyanobacteria; mixing of fecal matter; chemical contamination of water including infiltration of fertilizers, pesticides, chlorine, and other inorganic sources; and dissolved solid contamination of water comprising adulteration with heavy metals like lead, arsenic, cadmium, chromium, and iron. The natural or anthropogenic means of adulteration of drinking water institutes the foundation of water-associated infectious diseases which is the major global health problem.

Developing countries are the birthplace of water-associated infectious diseases as the basic necessities of water, hygiene (WASH), and sanitation are not according to the required standards. According to World Health Organization (WHO), 99.8% of deaths due to poor WASH conditions occur in developing countries where 90% of deaths are of children [3]. Additionally, 3.6% of the estimated global disease burden causing total disability-adjusted life years (DALYs) due to diarrheal diseases causing 1.5 million deaths every year [4]. The dominant reason behind such high toll is unsafe drinking water especially in low-income countries and being incompetent in practicing WASH [5].

Contaminated water is a safe haven for the germination of microbial pathogens. The changing source of water bodies changes the pathogens related to water-associated infectious diseases. In case of stagnant water, water-related infectious diseases are prevalent where mosquitoes are acting as vectors for malaria, dengue, Japanese encephalitis and lymphatic filariasis. Whereas, running water such as rivers are primarily responsible for water-based pathogens such as Onchocerciasis caused by *Onchocerca volvulus* (parasite) that breeds in water and is transmitted by black fly (*Simulium*), a vector that requires fast flowing water for its larval stages [6]. The presence of fecal origin pathogens in drinking water elucidates the improper processing and filtration of water supply which is another instance of running water-associated contamination of water. The irrigation of crops using fecal contaminated water in developing countries might be the root cause of water-related infectious diseases. Numerous water-washed and water-borne infectious diseases caused by

Table 7.1 Human pathogens transmitted from drinking water or recreational settings

Pathogens	Drinking water	Recreational settings
Pathogenic viruses		
Poliovirus	±	–
Coxsackie A viruses	–	–
Coxsackie B viruses	–	–
Echoviruses	–	–
Enteroviruses 68–71	–	–
Hepatitis A virus	+	+
Hepatitis E virus	+	–
Gastroenteritis viruses		
Norwalk-like viruses	+	+
Sapporo-like viruses	+	+
Astroviruses	+	–
Rotaviruses	+	–
Enteric adenoviruses 40, 41	–	–
Enteropathogenic bacteria		
Salmonella typhi	+	–
Salmonella paratyphi	+	–
Shigella spp.	+	+
Vibrio cholerae	+	+
Escherichia coli O157:H7 (VETC)	+	+
Yersinia enterocolitica	+	±
Campylobacter jejuni	+	+
Protozoa		
Acanthamoeba castellanii	–	+
Balantidium coli	+	–
Cryptosporidium hominis	+	+
Entamoeba histolytica	+	–
Giardia lamblia	+	–
Naegleria fowleri	+	–

contamination of drinking water with animal or human excreta are typhoid fever (*Salmonella typhi*), giardiasis (*Giardia lamblia*), cholera (*Vibrio cholerae*), poliomyelitis (polio viruses), gastroenteritis (rotaviruses), and amoebic dysentery (*Entamoeba histolytica*) [7]. Due to less salinity and alkaline in nature, fresh water bodies harbors fresh water snails which are the intermediate host for Schistosomiasis (Bilharzia). The eggs hatch releasing miracedia will generate from the *Schistosoma mansoni*/*Schistosoma haematobium*. Miracedia will then penetrate the snail tissues [8]. Human pathogens transmitted from drinking water or recreational settings have been highlighted in Table 7.1.

Various other reasons of contamination of water resources that eventually lead to rise in water-associated infectious diseases are traditional burial of dead body releasing decaying microbial agents in the soil that percolate during rainfall and adulterate the groundwater [9]. According to WHO, infections associated with dead bodies might be viral hemorrhagic fever (dengue) and gastrointestinal such as cholera, salmonellosis, and rotavirus diarrhea. Likewise, cyanobacterial eutrophication releases cyanotoxins that are ingested during recreational activity that leads to water-associated infectious diseases as reported by countries like China, Australia, the USA, Brazil, and South Africa [10].

References

1. Yang K, LeJeune J, Alsdorf D, Lu B, Shum CK, Liang S (2012) Global distribution of outbreaks of water associated infectious diseases. PLoS Negl Trop Dis 6(2):1–9
2. Racaniello VR (2004) Emerging infectious diseases. J Clin Invest 113(6):796–798
3. Uyttendaele M, Jaykus LA, Amoah P, Chiodini A, Cunliffe D, Jacxsens L, Holvoet K, Korsten L, Lau M, McClure P, Medema G (2015) Microbial hazards in irrigation water: standards, norms, and testing to manage use of water in fresh produce primary production. Compr Rev Food Sci Food Safety 14(4):336–356
4. World Health Organization (2012) Water sanitation hygiene. WHO, Geneva. http://www.who.int/water_sanitation_health/diseases-risks/en/. Accessed on 7-11-2017
5. Sorenson SB, Morssink C, Campos PA (2011) Safe access to safe water in low income countries: water fetching in current times. Soc Sci Med 72(9):1522–1526
6. Lakwo TL, Ndyomugyenyi R, Onapa AW, Twebaze C (2006) Transmission of Onchocerca volvulus and prospects for the elimination of its vector, the blackfly Simulium neavei in the Mpamba-Nkusi focus in Western Uganda. Med Vet Entomol 20(1):93–101
7. Leclerc H, Schwartzbrod L, Dei-Cas E (2002) Microbial agents associated with waterborne diseases. Crit Rev Microbiol 28(4):371–409
8. Fenwick A (2006) Waterborne infectious diseases—could they be consigned to history? Science 313(5790):1077–1081
9. Morgan O (2004) Infectious disease risks from dead bodies following natural disasters. Rev Panam Salud Publica 15(5):307–312
10. World Health Organization (2003) Emerging issues in water and infectious disease. WHO, Geneva. www.who.int/water_sanitation_health/emerging/emerging.pdf. Accessed on 6-11-2017

Global Strategies and Schemes for Preventing Water-Associated Infectious Diseases

8

Amrita Haikerwal*, Swatantra Kumar, and Shailendra K. Saxena*

Abstract

Water safety plans with a comprehensive analysis of risk management and risk assessment to provide safe and clean drinking water supply across the globe have been implemented. Global Hygiene and Drinking Water Analysis and Assessment (GLAAS) works with an objective to provide decision makers and policy makers a reliable, accessible, global analysis of the investments being enforced for providing the safety of drinking water, and maintaining hygiene and sanitation. WHO and UNICEF have initiated a Joint Monitoring Programme (JMP) that monitors the progress of WASH across all the countries. CDC has developed Safe Water System (SWS), a household water treatment system, that promotes behavioral changes to obtain a cost-effective and long-term solution.

Keywords

WSP · GLASS · JMP · WASH · Safe drinking water · Sanitation

8.1 Introduction

Leading health organizations such as the WHO, UN, and CDC have implemented robust, sustainable, and reliable strategies (Fig. 8.1) on a global front, providing the basic sanitation and water access for 2.3 billion people without privileged use [1].

---

*Contributed equally as first author

A. Haikerwal · S. Kumar · S. K. Saxena (✉)
Centre for Advanced Research, Faculty of Medicine, King George's Medical University, Lucknow, India
e-mail: shailen@kgmcindia.edu

S. K. Saxena (ed.), *Water-Associated Infectious Diseases*,
https://doi.org/10.1007/978-981-13-9197-2_8

Fig. 8.1 Global WASH implementation plans and strategies. Sustainable WASH management will be achieved through the implementation of global plans and strategies including Water Safety Plan, WHO-UNICEF JMP, and GLAAS

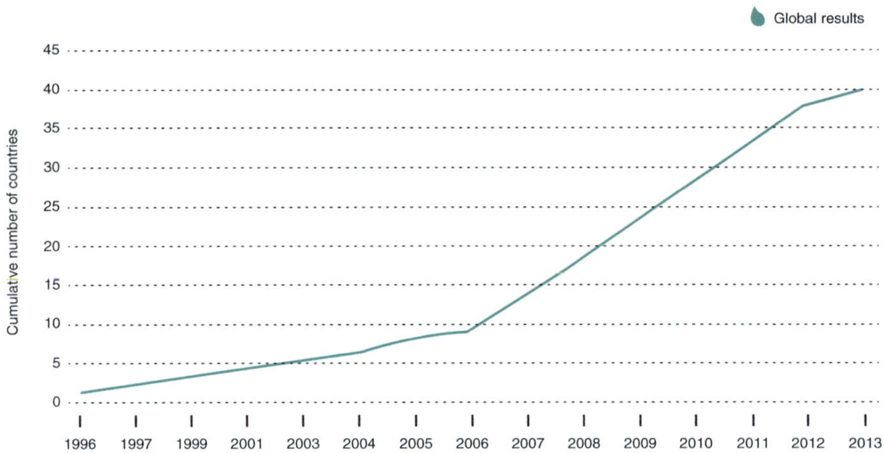

Fig. 8.2 Countries implementing WSPs ($n = 40$ respondent countries) [adapted from global status report on water safety plans: WHO]

8.2 Water Safety Plans (WSPs)

International Water Association (IWA) and WHO Bonn Charter for Safe Drinking Water in 2004 have introduced water safety plans with a comprehensive analysis of risk management and risk assessment to provide safe and clean drinking water supply across the globe, from the beginning of procuring water to delivering it to the consumers [2]. The implementation of WSP occurred in 93 countries around the globe (Fig. 8.2), where 72 countries have practiced WSP in both rural and urban regions.

8.3 Global Analysis and Assessment of Sanitation and Drinking Water (GLAAS)

A UN-Water proposal instigated by WHO [3] works with an objective to provide decision makers and policy makers a reliable, accessible, global analysis of the investments being enforced for providing the safety of drinking water, and maintaining hygiene and sanitation. The vision is to improve the status of WASH program as a sustainable development goal by monitoring the finances, stakeholders, governments, and sectors involving education and health, to minimize the gap and accelerate the implementation of WASH.

8.4 UNICEF/WHO Joint Monitoring Programme

WHO and UNICEF have initiated a condemnatory program (JMP) [4] that monitors the progress of WASH across all the countries. JMP majorly focuses on providing drinking water services where improved source of drinking water should be present on premises and is readily available with no source of fecal and chemical contaminations. They also aim to deliver sanitation services on the basis of some benchmarks where they want countries to be open defecation free, and maintain sanitation facilities that collect the human excreta and transport through sewer lines and treat them off-site. JMP aims to provide a universal hygiene service that provides accessibility to hand-washing facility. WHO has reported that 89% of world population is accessible to basic drinking water (Fig. 8.3) facility and 65% use basic sanitation services in 2015.

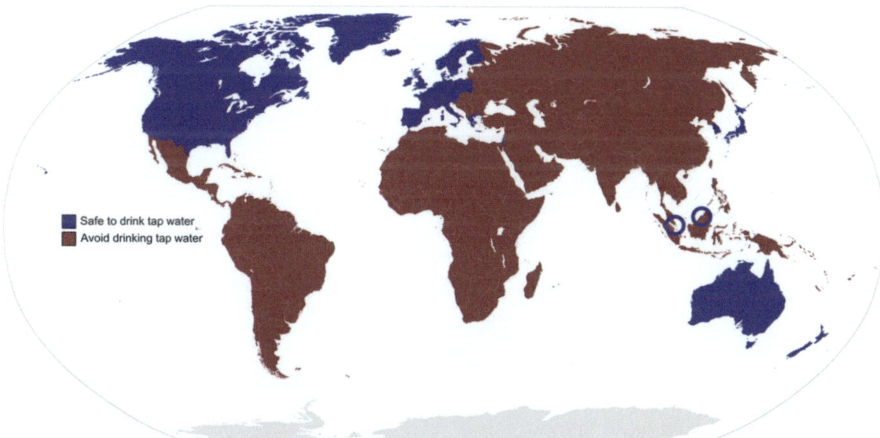

Fig. 8.3 The global scenario of safe drinking water in 2015 [adapted from JMP, WHO, and UNICEF]

8.5 CDC's Safe Water System

Zoonotic Infectious Diseases (NCEZID) and CDC's National Center for emerging water program ensure the safety of water and hygiene by regular monitoring and testing. CDC has developed Safe Water System (SWS), a household water treatment system, that promotes behavioral changes to obtain a cost-effective and long-term solution. Storage of treated water in households, healthcare facilities, and schools is the primary target in this program [4]. CDC assists UN in implementing hand-washing practices in communities, schools, and clinics. Furthermore, CDCs provide surveillance system, trained personnel, WASH expertise, and scientists to the site of water-associated infectious disease outbreak.

8.6 Swachh Bharat Mission: Open Defecation-Free India

A nationwide sanitation program, Swachh Bharat Mission (SBM), initiated on 14th October 2014 by the Prime Minister of India focuses to end open defecation by 2019. The government has promoted establishment of toilets in rural regions with an aim to provide a safe and healthy sanitation facilities (Fig. 8.4) to each and every one without any discrimination. According to the data reported on 8th November 2017 under SBM program 271,346 villages, 227 districts, and 7 states are open

Fig. 8.4 Swachh Bharat Mission reported the data showing current scenario of community and public toilets in India

defecation free [5]. The mission emphasizes on generation of awareness, changing perspective, and behavior of the communities towards adopting sustainable sanitation practices for generations. The monetary aid of $25 billion was granted in collaboration with the government, civil bodies, and the private sector.

References

1. World Health Organization (WHO) and the United Nations Children's Fund (UNICEF) (2017) Progress on drinking water, sanitation and hygiene: 2017 update and SDG baselines. http://www.who.int/mediacentre/news/releases/2017/launch-version-report-jmp-water-sanitation-hygiene.pdf. Accessed on 12-3-2018
2. World Health Organization Geneva, International Water Association (2017) Global status report on water safety plans: a review of proactive risk assessment and risk management practices to ensure the safety of drinking-water. http://www.who.int/water_sanitation_health/publications/global-status-report-on-water-safety-plans/en/. Accessed on 12-3-2018
3. World Health Organization Geneva, UN-Water (2017) UN-Water Global Analysis and Assessment of Sanitation and Drinking-Water (GLAAS) strategy 2017–2020. http://apps.who.int/iris/bitstream/10665/255649/1/WHO-FWC-WSH-17.03-eng.pdf?ua=1. Accessed on 12-3-2018
4. Centers for Disease Control and Prevention (CDC) (2013) National Center for Emerging and Zoonotic Infectious Diseases: our work, our stories, 2011–2012. CDC, Atlanta, GA. https://www.cdc.gov/ncezid/pdf/annual-report.pdf#page=90. Accessed on 12-3-2018
5. Government of India (2017) Swachh Bharat Mission—Gramin, Ministry of Drinking Water and Sanitation. Open defecation free. Swachh Bharat Mission. http://swachhbharatmission.gov.in/sbmcms/index.htm. Accessed on 8-11-2017

Novel Approaches for Detecting Water-Associated Pathogens

9

Vimal K. Maurya*, Swatantra Kumar, and Shailendra K. Saxena*

Abstract

Water-associated infectious diseases are the global concern for public health. In order to achieve the target of pathogen-free water, there is an urgent requirement to develop effective pathogen detection techniques. Advancement of pathogen indicators may help in the identification of large number of novel promising pathogens that cause water-related and water-borne diseases. Advancement of molecular techniques facilitates the differential diagnosis of various pathogens. The best recommended tests for the screening of water-associated pathogens are based on detection of the nucleic acid such as PCR, 16S rRNA, FISH, and DNA microarrays. Newer techniques are required which combine the strategy of accuracy assigned with traditional microbiology and sensitivity related with molecular biology. Biosensor technology has provided a platform which brings together both sensitivity and accuracy with rapid detection. Similarly, the early diagnosis of infectious diseases is the key to fight against it; thus, use of nano-diagnostics has proven to be promising in the detection of infectious agents as compared to traditional diagnostic methods.

Keywords

PCR · FISH · DNA microarrays · Biosensors · Nanotechnology · Indicators

*Contributed equally as first author

V. K. Maurya · S. Kumar · S. K. Saxena (✉)
Centre for Advanced Research, Faculty of Medicine, King George's Medical University, Lucknow, India
e-mail: shailen@kgmcindia.edu

S. K. Saxena (ed.), *Water-Associated Infectious Diseases*,
https://doi.org/10.1007/978-981-13-9197-2_9

9.1 Introduction

Water-borne infectious pathogens are the global apprehension for public health. The microbes present in water sources are the main cause of infection and severe illness in patients [1]. In order to achieve the target of pathogen-free water, there is an urgent requirement to develop effective pathogen detection techniques [2]. Even though available culture methods are tedious, require skilled person, and are less sensitive compared to other detection techniques. The major limitation of this method is that the viable but non-culturable microorganisms are not detected [3]. Advancement of molecular techniques facilitates the differential diagnosis of various pathogens. Improvement in available molecular technology can help to recognize causative microorganisms more precisely and in short duration [4]. Such techniques are also beneficial in the detection and characterization of viable microorganisms according to microbial communities which enables the available data that could increase the awareness of water-borne pathogens and protect public health [5]. Advancement of pathogen indicators may help in the identification of a large number of novel promising pathogens that cause water-related and water-borne diseases [6]. Quantitative microbial risk assessment (QMRA) is useful to measure the water safety and required for the measurement of risk from different pathogens and various types of microbial populations present in drinking water. QMRA is an effective approach to decreases the health risk by increasing the water quality [7]. The best recommended tests for the identification of water-associated pathogens are based on detection of the nucleic acid such as PCR, 16S rRNA, FISH, and DNA microarrays. PCR is the most frequently used technique for the recognition of water-associated pathogens. 16S rRNA is based on the detection of bacterial species in genetic library. Fluorescence in situ hybridization (FISH) has been found to be valuable for measuring the large quantity of microbial populations in any given sample. FISH is based on the principle of hybridization of rRNA probes tagged with fluorescent dye. The most recent molecular technique is DNA microarrays which can be used to test diverse water samples simultaneously [8]. Newer techniques are required which combine the strategy of accuracy assigned with traditional microbiology and sensitivity related with molecular biology. Biosensor technology has provided a platform which brings together both sensitivity and accuracy with rapid detection [9] (Fig. 9.1).

It also provides the opportunity of real-time monitoring of infectious agents present in the water bodies which allows the preventive measure to be compelled prior to any hazard in future. Label-free biosensors and peptide-based receptor molecule probes are the most effective strategies which are based on the various technologies of sensor platform [10]. Similarly, the early diagnosis of infectious diseases is the key to fight against it; thus, use of nano-diagnostics has proven to be promising in the detection of infectious agents as compared to traditional diagnostic methods [11]. Nano-diagnostic methods are sensitive and cost effective that might be provided in developing countries deprived of resources to fund high-cost molecular devices or a vaccination program for the protection of millions of people or that are dependent on international support [12]. However, nano-diagnostics could be used

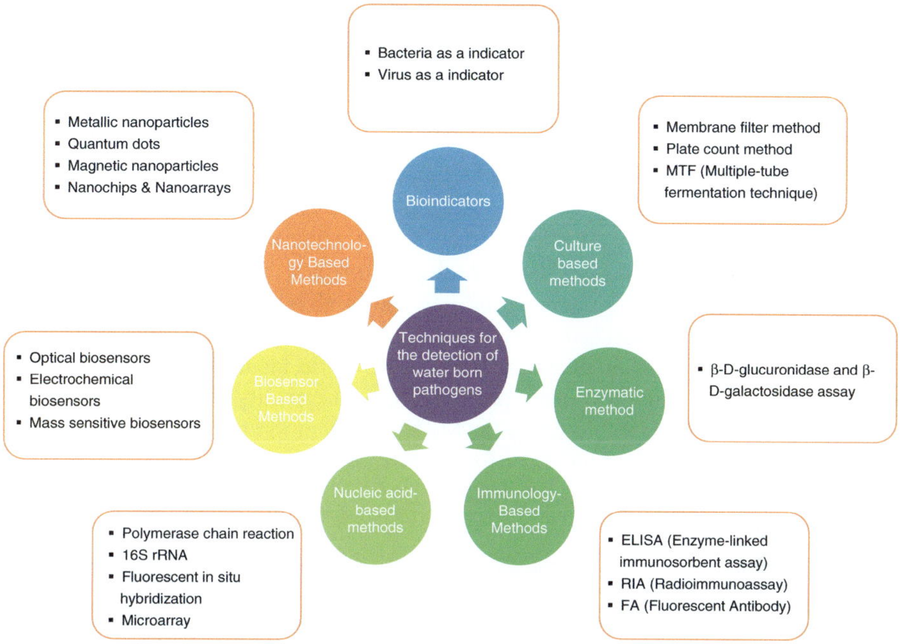

Fig. 9.1 Strategies for screening of water-associated pathogens

in remote areas as point-of-care unit for the diagnosis of infectious diseases. Numerous studies have reported the usage of nanoparticle base immunosensor for the rapid detection of infectious pathogens; for instance, *Vibrio cholerae* was detected by magnetite nanoparticles capped with citric acid, deposited onto glass substrate coated with indium tin oxide, and immobilized with monoclonal antibodies against *Vibrio cholerae* [13].

9.2 Indicators for Water-Associated Pathogens

Pathogens such as bacteria, virus, and protozoa that can be transmitted through drinking water are responsible for water-borne infectious diseases. Transmission of various pathogens in water bodies through pollution with human or animal feces is very high and it is very complex to screen the presence of specific pathogens [14]. The quality of water is mainly indicated by a certain species or group of microorganisms capable of measuring the full range of other microorganisms causing the risk to health. A potential water quality indicator must have some basic characteristics such as, it should be easily detectable and should not grow in natural water or general environment and should be present in large numbers in human and warm-blooded animal feces [15]. Currently, no bacterial indicator meets any ideal quality for measurement while various indicator organisms like enteric viruses for measurement of water quality are described with their advantages and limitations. Nowadays, indicator organisms such as total coliforms, *E. coli*, fecal coliforms, fecal streptococci, and enterococci are widely used for the screening of drinking water quality [16].

9.2.1 Total Coliforms and Fecal Coliforms

Total coliforms are commonly described as "Gram-negative, non-spore forming, all facultative anaerobic, oxidase-negative, β-galactosidase positive and rod-shaped bacteria that ferment lactose to acid and gas within 48 h at 35 °C" [17]. Coliforms are members of the family *Enterobacteriaceae*. Originally, the total bacterial coliform group was used as a substitute for *E. coli*, while identity of both (total and fecal) in raw water provides information about water quality. Two classical methods, named as most probable number (MPN) and membrane filtration (MF), are used for testing of total and fecal coliforms in water samples [18].

9.2.2 *E. coli*

E. coli belongs to a group known as thermotolerant fecal coliforms and is a specific indicator of fecal contamination present in water [19]. Existence of *E. coli* in food or water is not a direct indicator of pathogenic microorganisms but it indicates the higher risk of presence of other fecal borne bacteria and viruses such as *Salmonella* spp. or hepatitis A in the sample which are pathogenic in nature [20]. EPA (United States Environmental Protection Agency) recommends *E. coli* test to determine the level of enzyme which is selective for the *E. coli* organism in presumptive fecal coliforms. This test is used to differentiate the *E. coli* from non-fecal thermotolerant coliforms [21].

9.2.3 Fecal Streptococci and Enterococci

The various species of streptococci and enterococci are used as an indicator to determine the level of fecal contamination present in water resources [22]. There are six species of streptococci named as *S. faecalis*, *S. faecium*, *S. gallinarum*, *S. bovis*, *S. avium*, and *S. equines* that are used to measure the quality of water [23]. As an indicator, they possess various advantages over *E. coli* and coliform group such as they have longer survival, do not grow in the environment, and are easy to detect even after significant dilution. Streptococci and enterococci have a significant role during the assessment of water quality where they act as an additional indicator of treatment efficiency for detecting the contamination in both ground and surface water [24].

9.2.4 Bacteriophages

Bacteriophages are an alternative fecal indicator organisms that are better suited than fecal indicator bacteria (FIB) and also known as coliphages. Due to their closer biological and morphological properties they act as a surrogate for enteric viruses and other pathogens present in the natural environment [25]. In addition, bacteriophages are more suited to the indication of fecal contamination found in water resources than fecal bacteria because of their resistance to environmental factors. Somatic coliphages and F-specific RNA (F-RNA) phages are the two major categories of bacteriophages that are used in the detection of water quality. Phage detection of water samples involves several steps such as concentration of sample, decontamination of concentrate, and phage (plaque) assay by the double- or single-layer agar methods [26].

9.3 Identification of Potential Pathogens in Water Resources

9.3.1 Culture-Based Methods

Culture-based methods are considered as most precise and consistent techniques for the identification of water-borne pathogens. With the help of these methods it is possible to amplify the signal for the detection of single bacteria through the growth of a cell into a colony. All these procedures are time consuming because they require 72 h for the colony counts, about 18 h for the growth of a colony of 106 organisms, and different steps of serological confirmation, biochemical screening, selective enrichment, and pre-enrichment. Also the interpretation of results takes a long time and has a confirmation; often additional tests are required [27]. These methods are widely used for detection of *Salmonella*, *Campylobacter jejuni*, *Yersinia enterocolitica*, *S. aureus*, Coliforms, *E. coli*, *L. monocytogenes*, etc. [28]. The major problem of this technique is the time involved in the analysis: for example the detection of *L. monocytogenes* requires two steps of cultural enrichment that take 1 week only for the biochemical identification of the colonies and the detection of *Campylobacter* requires 4–9 days for a negative result and 14–16 days for the confirmation of a positive result [29]. Sometime it is possible to have an underestimation of pathogens because they can enter in a dormancy state, becoming non-culturable, and this is another point against this method. These techniques are not convenient in many industrial applications, particularly in food industry; for this reason they are often combined with other methods to obtain better results [30]. There are several traditional methods which are commonly used for the cultivation, enumeration, and detection of fecal indicators in the water resources like multiple-tube fermentation (MTF), membrane filtration method, and plate count enumeration method. Detection methods for pathogens and their detection limits have been mentioned in Table 9.1.

Table 9.1 Various detection methods for water-associated pathogens and their detection limits

Pathogen	Matrix sample	Detection method	Detection limit
E. coli	Contaminated tap water	Polymerase chain reaction (PCR)	1 cfu/100 mL
Enterotoxigenic E. coli (ETEC)	Water samples spiked by ETEC and nonpathogenic *E. coli*		4 cfu/mL
Cryptosporidium and *Giardia*	Environmental water samples		1–10 oocysts and 5–50 cysts
EHEC, Salmonella sp., *Shigella* sp., *Vibrio parahaemolyticus,* and *P. aeruginosa*	Polluted water and natural water	Multiplex PCR	10^1 cfu, 10^1 cfu, 10^2 cfu, 10^2 cfu, and 10^2 cfu
Adenovirus (adenovirus fiber gene in AdV40 and AdV41)	Wastewater, drinking water, recreational waters, and rivers	Quantitative polymerase chain reaction (qPCR)	5–8 copies of AdV40/41
L. monocytogenes, V. cholerae, V. parahaemolyticus, Pseudogulbenkiana sp., *S. flexneri, pathogenic E. coli, S. typhimurium*, and *C. perfringens*	Spiked environmental water samples (pond) and natural freshwater lake		10^2–10^4 cells per ca. 200 mg fecal samples of pathogens 100 cells/L
Enterovirus, adenovirus, Aichi virus, astrovirus, human hepatitis A and E viruses, norovirus, rotavirus, and sapovirus	River water, contaminated with effluents from a wastewater treatment plant		2 copies/μL of cDNA/DNA
V. cholerae	Ballast water	Real-time PCR	1 cfu/100 mL
C. wrairi, C. parvum, C. meleagridis, and *C. hominis*	Sewage and river water		1 oocyst
Astrovirus	Sewage		5–7 GC logs/100 mL
L. monocytogenes on biofilms	Artificial biofilms		6×10^2 cfu/cm^2
Bacillus anthracis, Brucella abortus, C. botulinum, Coxiella burnetii, Francisella tularensis, Rickettsia prowazekii, S. aureus, Yersinia pestis, Alexandrium catenella, Eastern equine encephalitis, *C. perfringens, V. cholerae, V. alginolyticus,* Ebola, Venezuelan equine encephalitis virus, and *Fusarium sporotrichioides*	Environmental water and ocean water spiked with pathogen	Microarrays	10 fg of *B. anthracis* DNA, 500 fg of *F. tularensis* DNA and *Y. pestis* DNA
S. enterica, Y. enterocolitica, E. coli, Typhimurium	Wastewater		1×10^7 *S. typhimurium* cells (10^4 *S. enteric* genomes)
Cryptosporidium, Acanthamoeba spp., *Blastocystis hominis, Entamoeba* spp., *Giardia intestinalis, Naegleria* spp.	Municipal wastewater treatment plants		1×10^3 target genes, or 50 *Cryptosporidium parvum* oocysts, per assay

Table 9.1 (continued)

Pathogen	Matrix sample	Detection method	Detection limit
Y. pestis	Milk	Pyrosequencing	0.9 cfu/mL
B. anthracis	Bottled water, milk, and juice		6 cfu/mL
Comamonadaceae, Proteobacteria, Bacteroidetes, Planctomycetes, and *Elusimicrobia*	Drinking water of the non-chlorinated distribution system		1.3×10^5 cells/mL
C. parvum	Oocysts diluted in PBS	Biosensors	10^5 oocysts/mL
E. coli 0157:H7 (NASBA + LFA biosensor)	Drinking water		40 cells/mL
V. cholerae (amperometric)	Ground and seawater		8 cfu/mL
Microcystis spp.	Lake water		30 ng/L
F. psychrophilum	Culture suspension	Fluorescence in situ hybridization (FISH)	7.3×10^5 cells/mL
E. coli (catalyzed reporter deposition (CARD)-FISH)	Mixed pure cultures and sludge		8.9 ± 1.5 16S rRNA molecules per cell
E. coli	Activated sludge		1400 ± 170 16S rRNA copies per *E. coli*
E. coli O157:H7 and *S. enterica typhimurium*	Contaminated food	Immunology-based methods	1.8×10^3 cfu/mL of *E. coli* and 9.2×10^3 cfu/mL of *Salmonella*

9.3.1.1 Multiple-Tube Fermentation Technique (MTF)

Most Probable Number (MPN) is used to calculate total microorganisms present in a given sample. The technique comprises inoculating a sequence of tubes with suitable decimal dilutions of the given sample [31]. After 48 h of incubation at 35 °C, formation of gas or acid or abundant growth in the test tubes confirms a positive reaction. All tubes with positive presumptive reaction are subsequently tested. The formation, at any time within 48 h at 35 °C, of gas in a brilliant green lactose bile broth tube provides positive confirmational testing [32].

9.3.1.2 Membrane Filter Method

A conventional method for screening the quality of drinking water is the membrane filter (MF). The technique of membrane filters is based on the filtering of water by means of a 0.45 mm pore-size sterile filter that adsorbs bacteria on the surface of the sterile filter. After filtering the water sample, filter is removed and incubated on selective medium (m-Endo-Type and Tergitol-TTC medium) and typical colonies on the filter are counted [33]. MF offers various advantages over MTF such as increased sensitivity, assurance, and testing of larger water volumes. MF is a widely used method in a majority of water quality laboratories [34].

9.3.1.3 Plate Count Method

Plate count is a conventional method used for assessment of viable pathogens in water or food source. Agar plates are used for dispersal of microorganisms and then incubated until the appearance of colonies; each colony represents single culturable microorganism or a clump of organisms. The method takes at least 24 h for the microorganisms to grow and form colonies of size that could be visualized for colony counting. However, many colonies may grow slow because of unfavorable cultivation media and temperature conditions. Therefore, most validated testing methods require an incubation period of at least 2–5 days [35].

9.3.2 Enzymatic Methods

The use of microbial enzymes is an attractive approach to detect the pathogens present in water sources. Depending upon the enzyme target, enzymatic reactions can be specific to the pathogen species, group, or genus [36]. β-D-Glucuronidase is a bacterial enzyme firstly identified in *E. coli* and catalyzes the hydrolysis of β-D-glucopyranosiduronic into D-glucuronic acid and aglycons [37]. Similarly there is another bacterial enzyme called β-D-galactosidase which is mostly used for counting the coliform group of family *Enterobacteriaceae* and also catalyzes the breakdown of lactose into glucose and galactose. These activities of both bacterial enzymes can be used for the detection and enumeration of *E. coli* and total coliforms in water samples [38]. Different chromogenic and fluorogenic substrates are used to detect specific enzymes in aquatic systems that generate color and fluorescence against the target in a given sample [39]. To identify the occurrence of β-D-glucuronidase in *E. coli*, various chromogenic substrates such as phenolphthalein-mono-β-D-glucuronide complex and indoxyl-β-D-glucuronide (IBDG) are used. For the detection of β-D-glucuronidase activity in *E. coli*, a fluorogenic substratum, such as 4-methylumbelliferyl β-D-glucuronide (MUGlu), is widely used. In the same way, β-D-galactosidase can also be detected in coliforms with 5-bromo-4-chloro-3-indolyl-β-D-galactopyranoside, *p* nitrophenyl-β-D-galactopyranoside, and *o*-nitrophenyl-β-D-galactopyranoside (chromogenic substrates) whereas for the detection of β-D-galactosidase activity in coliforms, 4-methylumbelliferyl-β-D-galactopyranoside (fluorogenic substrate) is used [40].

9.3.3 Molecular Methods

Molecular based methods are used to monitor a wide range of microorganisms present in a water sample and should be adapted by health authorities to enhance the quality of water [41]. Although every method has advantage and disadvantage that must be considered when understanding results molecular methods such as enzyme-linked immunosorbent assay (ELISA), fluorescent in situ hybridization (FISH), next-generation sequencing (NGS), polymerase chain reaction (PCR), 16S rRNA technology, microarrays, and pyrosequencing are more specific, selective, robust, and rapid methods compared to conventional or microscopic techniques when used for the screening of water-borne pathogens [42].

9.3.4 Immunology-Based Methods

They are based on antigen–antibody bindings that include heavy-chain, polyclonal, monoclonal, or recombinant antibodies for the identification of pathogens. These methods have also been widely applicable for the detection of toxins, spores, virus, and bacterial cells such as *Staphylococcus*, *Salmonella*, *L. monocytogenes*, *E. coli*, and *Campylobacter*. The immunological recognition of pathogens has been increased after the development of monoclonal antibodies being more specific, more sensitive, and reliable due to availability of large amount of single antibodies [43]. Immunological methods such as enzyme-linked immunosorbent assay (ELISA), serotyping, fluorescent antibody (FA), radioimmunoassay (RIA), and immunomagnetic separation are widely used for the detection of microbes present in food or water samples. They are more robust, faster and able to detect not just organisms which contaminate, but also their biotoxins which cannot be expressed in the organism's genome. They are not just as specific or sensitive as nucleic acid detection. The problems associated are the potential interference of contaminants, low sensitivity, and low affinity between antibody and pathogen or with other analytes measured [44].

9.3.5 Nucleic Acid-Based Methods

9.3.5.1 Polymerase Chain Reaction (PCR)

PCR is a common laboratory technique realized about 20 years ago and can be used to amplify small amount of genetic material to determine the presence of pathogens [45]. The PCR method is quite delicate; a single pathogenic bacterium in water samples can actually be detected, because in less than an hour a target sequence of DNA can be amplified 1 million times. This method is used to improve the sensitivity of nucleic acid-based methods due to its low percentage of false positives [46]. PCR has an advantage in relation to other methods; with the help of this technique we can detect very small amounts of target nucleic acid present in a given sample. Other advantages such as high sensitivity, selectivity, potential for automation, specificity, and rapid detection with high accuracy make this technique a valuable alternative among various industries [47]. This technique has been useful for the identification of pathogens present in water or food samples like *Salmonella*, *S. aureus*, *Bacillus cereus*, *C. jejuni*, *L. monocytogenes*, *Yersinia enterocolitica*, and *Escherichia coli* O157:H7 [48]. Real-time PCR, multiplex PCR, and reverse transcriptase PCR (RT-PCR) are the special PCR-based methods for detection of pathogen. In the last years there has been a development in PCR techniques, as reported in literature with multiplex PCR that permitted to detect simultaneously several pathogens with the use of various primers that amplify DNA regions coding for specific genes of each targeted microbial strain [49]. PCR has a significant disadvantage, since the DNA is always present in cells which are dead or alive and cannot be distinguished between viable and nonviable cells. The reverse transcriptase PCR (RT-PCR) can overcome this limitation [50]. The principal issue in the detection of microbes using PCR is that it is expensive, complicated to use, and requires pure samples, time for processing, expertise in molecular biology, and skilled workers. From all these assays we can deduct how today there is the need of a new device for an in situ, simple, inexpensive, and quick detection that does not require skilled workers, with high sensitivity, for a fast screening of different samples and that gives an immediate response [51].

9.3.5.2 16S rRNA

16S ribosomal RNA sequencing technique is based on identification of universal and abundant RNA of the small ribosomal subunit. This technique not just recognizes the microorganism; it additionally distinguishes the microbial species present in the sample mixture and gives a genetic library for microbial masses in any certain sample [52]. As a result, this technique has great importance in various fields especially in microbial ecology where this technique is used to provide accurate taxonomic data, identify pathogens that have not yet been cultured, and detect small microbial groups present in the given sample [53]. 16S rRNA approach begins with the extraction of nucleic acid followed by PCR amplification using 16S rRNA universal primers, after that cloning, sequencing, identification, and genetic library construction [54]. The sequence of 16S rRNA genes is enough to produce taxonomic information that can be analyzed in a phylogenetic tree [55]. Additionally, this technique is used as an indicator to pick up the dominating microorganism in the water sample. Other molecular strategies like terminal restriction fragment length polymorphism (tRFLP), 16S-RFLP (ARDRA), and ribosomal intergenic spacer analysis (RISA) are based on the amplification and evaluation of the rRNA gene [56]. As for any technique, 16S rRNA approach has some limitations like this technique is not suitable for the monitoring of viruses [57].

9.3.5.3 Fluorescent In Situ Hybridization

FISH is useful to measure the loads in any microbial sample of the respective populations. This method also specifies microbial cells along with the generation of quantitative data of the microbial population in a sample [58]. FISH starts with treating the microbial cells with a fixative solution followed by hybridizing the cells on a glass slide with particular probes and then visualizing these probes with the help of confocal laser microscopy or epifluorescence [59]. Further steps for the sample preparation are homogenization, pre-enrichment procedures, or bacterial separation. The conjugation of FISH with confocal microscopy has turned into a potential observing methodology in wastewater treatment plants.

9.3.5.4 Microarray

Microarray technology is a molecular biology-based one cutting-edge approach. The development of the microarray technology enables health professionals to simultaneously carry out massive, data-intensive experiments on environmental water samples [60]. Microarray technology consists of hybridization to a complementary sequence (i.e., the probe) of a target sequence of nucleic acids (DNA or RNA). Microarray samples generally target taxonomic genes like 16S or functional genes [61]. Electrochemical detection (ECD) 12K microarray is the most advanced microarray technology based on semiconductor matrix that has more than 12,000 nucleic acid probes designed on a single chip. DNA microarrays have been successfully used for investigating *E. histolytica* pathogenesis [62]. An oligonucleotide microarray was developed for simultaneous detection and genotyping of protozoan parasites causing water-borne disease outbreaks, *E. histolytica*, *C. parvum*, *G. duodenalis*, and *E. dispar*. It's particularly interesting to use species-specific DNA samples to hybridize unamplified DNA isolated from fecal samples, especially when using radioactive labels. Dot blot was found to be more sensitive as compared to microscopy and isoenzyme analysis. The method has been successfully applied with probes of different genomic regions from rDNA and tandem repeat probe but has not been widely employed, because it may cross hybridize sometimes [63].

9.3.5.5 Next-Generation Sequencing

The recent emergence of novel techniques such as next-generation sequencing (NGS) has started to revolutionize the infectious disease diagnosis and become an attractive strategy for directly sensing microbial sequences in water samples without previous awareness of the type of contagion [64]. NGS provides the complete genome sequence of currently known microorganisms and this information is utilized for construction of phylogenetic tree as well as detection of pathogens. This technique may report unique sequences of isolates from the sources and by counting the reads of each sequence provides important information about the contamination load in each water sample [65]. NGS also provides important and useful signature sequences of pathogens present in a given sample. Different studies have identified viruses among the top metagenomics applicants because they have a unique nuclear acid—capsid protein coding (i.e., hardened capsids) to sustain the sample treatment procedure in the virus concentration and purification laboratories [66]. The genome size of viruses, mainly bacteriophages, has been beneficial for bioinformatics and genome assembly. Furthermore, viruses have no common diagnostic marker like the bacterial 16S rRNA gene [67]. For that reason, NGS is able to accommodate millions of short metagenomic sequences for contiguous assembly to detect viral members in a given sample population [68].

9.3.5.6 Pyrosequencing

Pyrosequencing is a molecular level sequencing method used to screen microbial metagenomes in a wide range of environmental samples including reused wastewater, freshwater, marine water, and open ocean water and in real time provides a huge number of sequences read in a single run [69]. In pyrosequencing, enzymes such as luciferase, DNA polymerase, apyrase, and ATP sulfurylase are required for the detection of pathogens [70] and at some stage in the nucleotide integration the pyrophosphate (PPi) is released. The light signal which is produced during the conversion of PPi into ATP in the presence of ATP sulfurylase and finally ATP to visible light by enzyme luciferase is detected [71].

9.3.6 Biosensor-Based Methods

Biosensor-based methods are commonly used for screening of water-borne pathogens and environmental toxins. Conventional pathogen detection methods often take time and are complicated. Therefore, a new, reliable, sensitive, fast, specific, and straightforward low-cost technology needs to be developed to monitor pathogens in a given sample in real time [72]. The design of biosensor is a fusion between molecular biology and information technology that together give information about the interaction of biomolecule-analyte [73] (Fig. 9.2).

In biosensors, the conversion of a biological response into an electrical signal is possible with the help of two main elements: a bioreceptor which identifies the target and a transducer that converts the recognition in an electrical signal [74]. The bioreceptor used for detection of pathogens is composed of highly specialized macromolecules or complex systems with an appropriate selectivity and sensitivity [75]. The transducer, which is in close contact with the bioreceptor, determines different properties or the optical activity of the target samples [76].

9.3.6.1 Bioreceptors

Bioreceptors are the elements that give specificity to the biosensor through a biochemical mechanism. Binding with the analyte under investigation produces a physicochemical effect that can be detected by the transducer [77]. Bioreceptors can be classified into seven categories: enzymes, proteins, peptides, biomimetics and bacteriophages, antigen–antibody, cells, and nucleic acids. The most used bioreceptors are nucleic acids and enzymes [78].

9.3.6.2 Transducers

Transducer is a major factor in detecting pathogens associated with water because they show that the bioreceptor reaction has taken place. Transduction can be proficient with a great variety of methods. The choice of the transducer depends on the reaction type, on the substances liberated or consumed, and on the application of the biosensor [79]. Various types of transducers are available but the most common and popular are optical based transducers, electrochemical based transducers, and mass-based transducers [80].

Optical Biosensors

Optical biosensors have been used to identify pathogens and environmental toxins due to their high selectiveness and sensitivity [81]. The various types of spectroscopic methods such as infrared, Raman, SERS, reflection, refraction, dispersion,

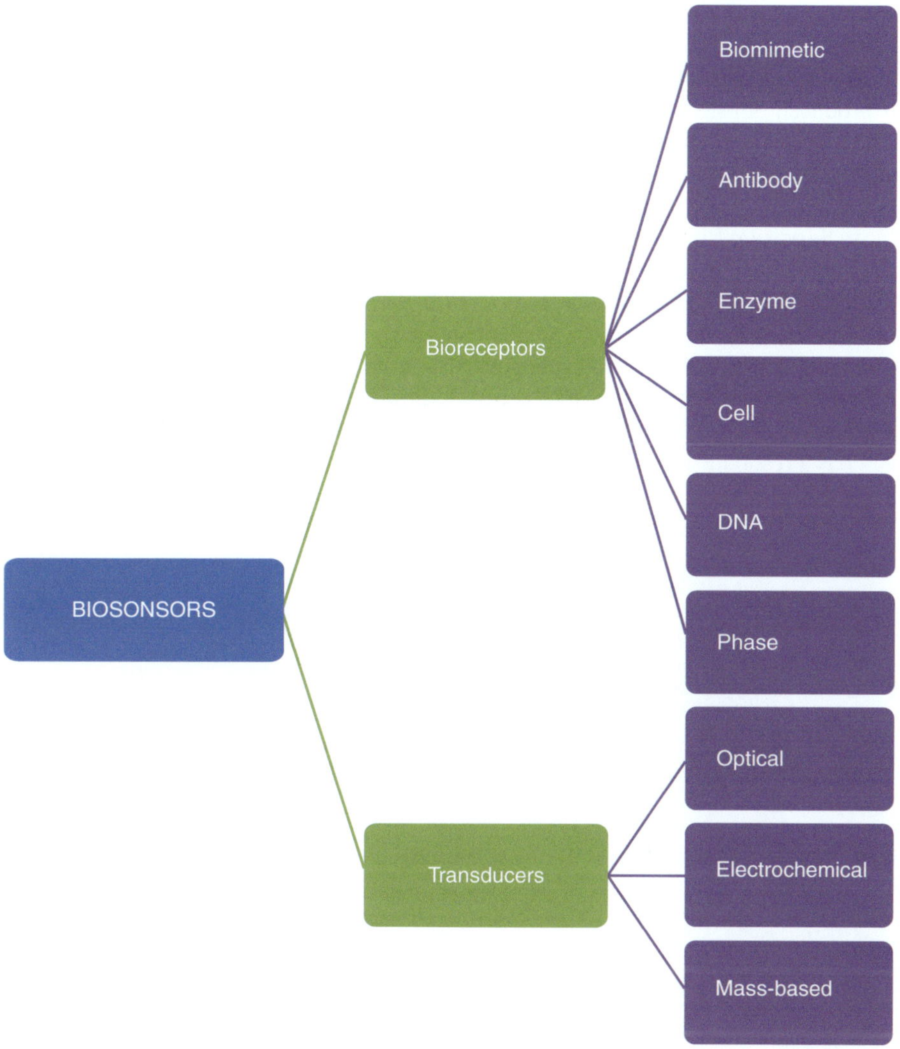

Fig. 9.2 Classification of biosensors

absorption, fluorescence, chemiluminescence, and phosphorescence are used in optical based methods to determine the characteristics of sample. These biosensors are capable of sensing even minor changes of analytes, thus allowing sensitive and specific measurements. Optical biosensor is used for the detection of fluorescent-labeled *E. coli* O157:H7 where the limit of detection was 360 cell/mL.

Electrochemical Biosensors

Recently, various electrochemical biosensors are developed and anticipated for screening of various diseases. The detection is based on specific biomarkers with improvement in biosensor characteristics including rapid response, selectivity, low cost, and sensitivity [82]. They can be further classified into conductometric, potentiometric, amperometric, and impedimetric. This classification depends on the measurement of changes in current, conductance, voltage, and impedance, respectively, which is produced during the interaction of analyte and bioreceptor [83]. The current produced by the oxidation and reduction reactions is directly related with concentration of the electroactive species present, and is determined by these kinds of biosensors.

Mass Sensitive Biosensors

They have been used to determine the small changes in mass during the interaction of highly specialized macromolecules. Piezoelectric sensors and surface plasmon resonance (SPR)-based sensors are the commonly used mass biosensors for the identification of water-associated pathogens. The quartz crystal microbalance biosensor is a mass-sensitive biosensor that is inserted in two electrodes using quartz crystal. As quartz is piezoelectric; it can be excited by applying a voltage across the elements and has a resonance frequency. It is used for wave propagation [84] (Fig. 9.3).

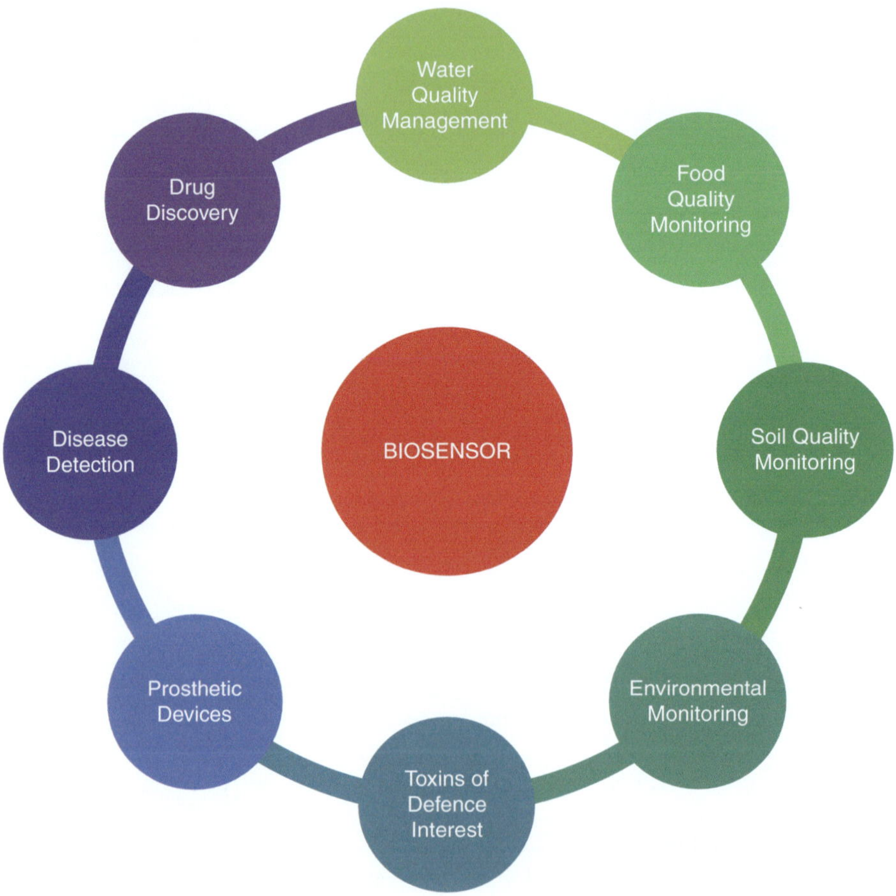

Fig. 9.3 Major areas of applications for biosensors

9.3.7 Nanotechnology-Based Methods

Nanotechnology helps to design specific, selective, and economical diagnostic means for the screening of water-associated pathogens [85]. In the last years, nanomaterials have been considered really interesting in the field of nanoscience and various nanostructures have been studied as nanorods, nanotubes, nanoparticles, nanofibers, and thin films and used for different applications [86]. Pathogens present in water and food samples can be detected with the help of nanoscale materials which are coupled with various types of sensors. Due to their size and composition, metal nanoparticles such as gold or silver contain several optical and electronic properties [87]. When these properties are coupled with affinity ligands, they can be used for the screening of various pathogens present in water or food samples [88]. Various nanotechnology-based approaches such as carbon nanotubes, metallic nanoparticles, and quantum dots when conjugated with specific oligonucleotides can identify complementary DNA strands which are detected by color changes and used in DNA detection and step-up of immunoassays for various water-associated pathogens [89]. Recently, there are several studies for the development of sensors for the identification of pathogenic microorganisms and environmental toxins. The need for rapid and sensitive testing methods for detecting pathogens from water and environmental toxins has led to biosensor technology being incorporated into microarray and other platforms [90] (Fig. 9.4).

To improve the detection methods it has been necessary to follow new routes as nanotechnologies, to develop nanosensors for the detection of any kind of traces of contaminants [91]. The possibility of combining biology and nanoscale technology into sensors and the incorporation of nanoelements in various procedures decreases the dimensions of the instruments and increases the automation, sensitivity, and rapidity of results [92]. Because of their specificity, high sensitivity, and ability to operate without enzymes, several techniques for medical and medical testing were developed based on nanomaterials [93]. A variety of nanomaterial such as metallic nanoparticles, fluorescent quantum dots, magnetic nanoparticles, fluorescent polymeric nanoparticles, fiber-optic-based biosensors, cantilever-based arrays, nanochips, and nanoarrays have been developed in the sandwich-type immunosensors and immunoassays [94]. List of nanoparticles studied in the diagnosis of water-associated infections has been highlighted in Table 9.2.

9.3.7.1 Metallic Nanoparticles

In several diagnostic devices for the detection of pathogens, chemicals, and toxin, metal nanoparticles like gold or silver nanoparticles were widely used for signal amplification [95]. Gold nanoparticles (Au NPs) in conjugation with optical and electrical biosensors are used for the monitoring of multiple drug-resistant *S. typhimurium* DT104 and *S. typhimurium* [96, 97]. Similarly, silver nanoparticles (Ag NPs) are conjugated with ligands like oligonucleotides, proteins, peptides, and antibodies. These nanoparticles can also be of help in microscopic identification of targets using both electrochemical approaches and scanning transmission electron microscopy (STEM) [98].

Fig. 9.4 Nanomaterials used for the pathogen detection

9.3.7.2 Quantum Dots

Quantum dots are the perfect tool for various biomedical applications, such as a sensor and biomarker detection tool, pathogens, cells, and tissue immunolization. QDs have also proven to be promising microbial detection nanomaterials [99]. QDs are linked to specific organisms, such as the parasites (*Cryptosporidium parvum* and *Giardia lamblia*) and bacterial species (*Listeria monocytogenes*, *Escherichia coli* O157:H7, *Mycobacterium bovis*, *Shigella*, and *Salmonella*) to facilitate the detection of organisms [100].

9.3.7.3 Magnetic Nanoparticles

The role of magnetic NPs is limited in the monitoring of pathogens in water or food samples. However, they have been used solely to separate, purify, or concentrate different biomolecules, including those from microbes or complex matrices [101].

Table 9.2 List of nanoparticles studied in the diagnosis of water-associated infections

Nanomaterial	Pathogen	Detection method	Recognition	Detection limit
Qdots	*E. coli* O157:H7	Fluorescence microscopy	Biotinylated antibody	Two orders more sensitive than conventional dyes
Magnetic NPs	*E. coli* O157:H7	Fluorometry visual	Biotinylated antibody	100 times more sensitive than FITC 1680 cells
Au NPs	*M. tuberculosis*	UV–vis spectroscopy	MycoD1 Oligonucleotide	Visual detection with minimum detection of 18.75 ng of mycobacterial DNA
Influenza-specific antibody-conjugated AuNPS	Influenza virus	Dynamic light scattering	Influenza virus	b100 TCID50/mL
Titanium NPs	*S. aureus*	MALDI-MS	Pathogen	Detection without the requirement of any culturing steps or any other sample pretreatment techniques
Au NPs/Ag staining	HBV, HCV antibodies	Protein chip assay	Protein A	Visual detection

Magnetometers or superconducting quantum interference devices (SQUIDs) are used to detect changes in the magnetic properties of particles while interacting with specific targets in a molecular environment [102]. They have been used extensively to identify different targets like mRNA, DNA, bacteria, and viruses. Their implementation in biosensors is an important strategy for the screening of microorganisms present in various environmental samples [103].

9.3.7.4 Nanochips and Nanoarrays

Recently, nanochips are used for the rapid identification of microbes present in food or water sample. In conjugation with gold or silica nanoparticles, nanochips can also be used for the monitoring of biomolecules including toxins and antibodies. The high-throughput sample screening efficiency depends on the type of currently available instrumentation and size of nanochips used for the detection [85]. The nanoarray can use different samples, including antibodies, nucleic acid, and small molecules, which can be deposited either in droplets of 1–30 μm or through fluidic molecular transfer operations [104].

9.4 Conclusions and Future Perspectives

Water-borne infectious diseases have become known as a global health issue. A promising diagnostic method can contribute to reducing disease transmission by rapid and sensitive detection by environmental water samples. An ideal detection method needs to accomplish the listed requirements such as high specificity and sensitivity, short time to results, great operational simplicity, and cost-effectiveness. Advancement of molecular techniques facilitates the differential diagnosis of various pathogens. Improvement in available molecular technology can help to identify causative agents more precisely and in a shorter duration. Different methods for rapid identification and quantification of water-borne pathogens in water have been developed and classified into immunology-based, nucleic acid-based, biosensor-based, and nanotechnology-based methods. The best recommended tests for the detection of water-associated pathogens are based on the detection of the nucleic acid such as PCR, 16S rRNA, FISH, and DNA microarrays. Advancement of pathogen indicators may help in the identification of a large number of novel promising pathogens that cause water-related and water-borne diseases. The application of quantitative microbial risk assessment (QMRA) is needed for the measurement of risk from different pathogens and type of microbial populations in drinking water which decreases the health risk by increasing the water quality.

References

1. Bentham R, Whiley H (2018) Quantitative microbial risk assessment and opportunist waterborne infections—are there too many gaps to fill? Int J Environ Res Public Health 15(6):E1150
2. Párraga-Niño N, Quero S, Ventós-Alfonso A, Uria N, Castillo-Fernandez O, Ezenarro JJ, Muñoz FX, Garcia-Nuñez M, Sabrià M (2018) New system for the detection of Legionella pneumophila in water samples. Talanta 189:324–331
3. Doan T, Pinsky BA (2016) Current and future molecular diagnostics for ocular infectious diseases. Curr Opin Ophthalmol 27(6):561–567
4. Alhamlan FS, Al-Qahtani AA, Al-Ahdal MN (2015) Recommended advanced techniques for waterborne pathogen detection in developing countries. J Infect Dev Ctries 9(2):128–135
5. Leclerc H, Schwartzbrod L, Dei-Cas E (2002) Microbial agents associated with waterborne diseases. Crit Rev Microbiol 28(4):371–409
6. Ahmed AHS, Aaron ME, Alison O, Claudia VL, Matthew KL (2016) How do ecologists select and use indicator species to monitor ecological change? Insights from 14 years of publication in ecological indicators. Ecol Indic 60:223–230
7. Schijven JF, Teunis PF, Rutjes SA, Bouwknegt M, de Roda Husman AM (2011) QMRAspot: a tool for quantitative microbial risk assessment from surface water to potable water. Water Res 45(17):5564–5576
8. Girones R, Ferrús MA, Alonso JL, Rodriguez-Manzano J, Calgua B, Corrêa Ade A, Hundesa A, Carratala A, Bofill-Mas S (2010) Molecular detection of pathogens in water—the pros and cons of molecular techniques. Water Res 44(15):4325–4339
9. Du X, Zhou J (2018) Application of biosensors to detection of epidemic diseases in animals. Res Vet Sci 118:444–448
10. Campuzano S, Yáñez-Sedeño P, Pingarrón JM (2017) Molecular biosensors for electrochemical detection of infectious pathogens in liquid biopsies: current trends and challenges. Sensors (Basel). 17(11):E2533
11. Wang Y, Yu L, Kong X, Sun L (2017) Application of nanodiagnostics in point-of-care tests for infectious diseases. Int J Nanomed 12:4789–4803
12. Ikoba U, Peng H, Li H, Miller C, Yu C, Wang Q (2015) Nanocarriers in therapy of infectious and inflammatory diseases. Nanoscale 7(10):4291–4305

13. Saunders NA (2011) Application of nanomaterials to arrays for infectious disease diagnosis. Nanomedicine (Lond) 6(2):271–280
14. Farkas A, Drăgan-Bularda M, Ciatarâş D, Bocoş B, Tigan S (2012) Opportunistic pathogens and faecal indicators in drinking water associated biofilms in Cluj, Romania. J Water Health 10(3):471–483
15. Won G, Gill A, Lejeune JT (2013) Microbial quality and bacteria pathogens in private wells used for drinking water in northeastern Ohio. J Water Health 11(3):555–562
16. Ferguson AS, Layton AC, Mailloux BJ, Culligan PJ, Williams DE, Smartt AE, Sayler GS, Feighery J, McKay LD, Knappett PS, Alexandrova E, Arbit T, Emch M, Escamilla V, Ahmed KM, Alam MJ, Streatfield PK, Yunus M, van Geen A (2012) Comparison of fecal indicators with pathogenic bacteria and rotavirus in groundwater. Sci Total Environ 431:314–322
17. Rompré A, Servais P, Baudart J, de-Roubin MR, Laurent P (2002) Detection and enumeration of coliforms in drinking water: current methods and emerging approaches. J Microbiol Methods 49(1):31–54
18. Maheux AF, Boudreau DK, Bisson MA, Dion-Dupont V, Bouchard S, Nkuranga M, Bergeron MG, Rodriguez MJ (2014) Molecular method for detection of total coliforms in drinking water samples. Appl Environ Microbiol 80(14):4074–4084
19. Tallon P, Magajna B, Lofranco C, Leung KT (2005) Microbial indicators of faecal contamination in water: a current perspective. Water Air Soil Pollut 166:139–166
20. Lin J, Ganesh A (2013) Water quality indicators: bacteria, coliphages, enteric viruses. Int J Environ Health Res 23(6):484–506
21. Reynnells R, Ingram DT, Roberts C, Stonebraker R, Handy ET, Felton G, Vinyard BT, Millner PD, Sharma M (2014) Comparison of U.S. Environmental Protection Agency and U.S. Composting Council microbial detection methods in finished compost and regrowth potential of Salmonella spp. and Escherichia coli O157:H7 in finished compost. Foodborne Pathog Dis 11(7):555–567
22. Althaus H, Dott W, Havemeister G, Müller HE, Sacré C (1982) Faecal streptococci as indicator organisms of drinking water. Zentralbl Bakteriol Mikrobiol Hyg A 252(2):154–165
23. Stella EI, Ifeanyi OE (2018) A review on salmonella species and indicator organisms in drinking water. Int J Compr Res Biol Sci 5(2):5–23
24. Briancesco R (2005) Microbial indicators and fresh water quality assessment. Ann Ist Super Sanita 41(3):353–358
25. Méndez J, Audicana A, Cancer M, Isern A, Llaneza J, Moreno B, Navarro M, Tarancón ML, Valero F, Ribas F, Jofre J, Lucena F (2004) Assessment of drinking water quality using indicator bacteria and bacteriophages. J Water Health 2(3):201–214
26. Hartard C, Banas S, Loutreul J, Rincé A, Benoit F, Boudaud N, Gantzer C (2016) Relevance of F-specific RNA bacteriophages in assessing human norovirus risk in shellfish and environmental waters. Appl Environ Microbiol 82(18):5709–5719
27. Larsen LH, Lange J, Xu Y, Schønheyder HC (2012) Optimizing culture methods for diagnosis of prosthetic joint infections: a summary of modifications and improvements reported since 1995. J Med Microbiol 61(Pt 3):309–316
28. Sanders SQ, Boothe DH, Frank JF, Arnold JW (2007) Culture and detection of Campylobacter jejuni within mixed microbial populations of biofilms on stainless steel. J Food Prot 70(6):1379–1385
29. Murphy NM, McLauchlin J, Ohai C, Grant KA (2007) Construction and evaluation of a microbiological positive process internal control for PCR-based examination of food samples for Listeria monocytogenes and Salmonella enterica. Int J Food Microbiol 120(1–2):110–119
30. Iqbal SS, Mayo MW, Bruno JG, Bronk BV, Batt CA, Chambers JP (2000) A review of molecular recognition technologies for detection of biological threat agents. Biosens Bioelectron 15(11–12):549–578
31. Kuai L, Nair AA, Polz MF (2001) Rapid and simple method for the most-probable-number estimation of arsenic-reducing bacteria. Appl Environ Microbiol 67(7):3168–3173

32. Wohlsen T, Bates J, Vesey G, Robinson WA, Katouli M (2006) Evaluation of the methods for enumerating coliform bacteria from water samples using precise reference standards. Lett Appl Microbiol 42(4):350–356
33. Liu H, Yang Y, Cui J, Liu L, Liu H, Hu G, Shi Y, Li J (2013) Evaluation and implementation of a membrane filter method for Cronobacter detection in drinking water. FEMS Microbiol Lett 344(1):60–68
34. Hay J, Khan W, Mead AJ, Seal DV, Sugden JK (1994) Membrane filtration method for bacteriological testing of water: enhanced colony visualization and stability on purification of phenol red indicator. Lett Appl Microbiol 18(2):117–119
35. Ekawati ER, Yusmiati SN (2018) Detection of Salmonella sp., Vibrio sp. and total plate count bacteria on blood cockle (Anadara granosa). In: IOP conference series: earth and environmental science
36. Stadler P, Loken LC, Crawford JT, Schramm PJ, Sorsa K, Kuhn C, Savio D, Striegl RG, Butman D, Stanley EH, Farnleitner AH, Zessner M (2018) Spatial patterns of enzymatic activity in large water bodies: ship-borne measurements of beta-D-glucuronidase activity as a rapid indicator of microbial water quality. Sci Total Environ 651(Pt 2):1742–1752
37. George I, Petit M, Servais P (2000) Use of enzymatic methods for rapid enumeration of coliforms in freshwaters. J Appl Microbiol 88(3):404–413
38. Caruso G, Crisafi E, Mancuso M (2002) Development of an enzyme assay for rapid assessment of Escherichia coli in seawaters. J Appl Microbiol 93(4):548–556
39. Perry JD, James AL, Morris KA, Oliver M, Chilvers KF, Reed RH, Gould FK (2006) Evaluation of novel fluorogenic substrates for the detection of glycosidases in Escherichia coli and enterococci. J Appl Microbiol 101(5):977–985
40. Ramsay J (2013) High-throughput ß-galactosidase and ß-glucuronidase assays using fluorogenic substrates. Bio-protocol 3(14):e827
41. Straub TM, Chandler DP (2003) Towards a unified system for detecting waterborne pathogens. J Microbiol Methods 53(2):185–197
42. Ramírez-Castillo FY, Loera-Muro A, Jacques M, Garneau P, Avelar-González FJ, Harel J, Guerrero-Barrera AL (2015) Waterborne pathogens: detection methods and challenges. Pathogens 4(2):307–334
43. Amani J, Mirhosseini SA, Imani Fooladi AA (2014) A review approaches to identify enteric bacterial pathogens. Jundishapur J Microbiol 8(2):e17473
44. Lim DV, Simpson JM, Kearns EA, Kramer MF (2005) Current and developing technologies for monitoring agents of bioterrorism and biowarfare. Clin Microbiol Rev 18(4):583–607
45. Law JW, Ab Mutalib NS, Chan KG, Lee LH (2015) Rapid methods for the detection of foodborne bacterial pathogens: principles, applications, advantages and limitations. Front Microbiol 5:770
46. Cordray MS, Richards-Kortum RR (2012) Emerging nucleic acid-based tests for point-of-care detection of malaria. Am J Trop Med Hyg 87(2):223–230
47. Valones MA, Guimarães RL, Brandão LA, de Souza PR, de Albuquerque Tavares Carvalho A, Crovela S (2009) Principles and applications of polymerase chain reaction in medical diagnostic fields: a review. Braz J Microbiol 40(1):1–11
48. Deshmukh RA, Joshi K, Bhand S, Roy U (2016) Recent developments in detection and enumeration of waterborne bacteria: a retrospective minireview. Microbiology 5(6):901–922
49. Touron A, Berthe T, Pawlak B, Petit F (2005) Detection of Salmonella in environmental water and sediment by a nested-multiplex polymerase chain reaction assay. Res Microbiol 156(4):541–553
50. Zeng D, Chen Z, Jiang Y, Xue F, Li B (2016) Advances and challenges in viability detection of foodborne pathogens. Front Microbiol 7:1833
51. Mirmajlessi SM, Destefanis M, Gottsberger RA, Mänd M, Loit E (2015) PCR-based specific techniques used for detecting the most important pathogens on strawberry: a systematic review. Syst Rev 4:9
52. Ward DM, Weller R, Bateson MM (1990) 16S rRNA sequences reveals numerous uncultured microorganisms in a natural community. Nature 345(6270):63–65

53. Fredricks DN (2001) Microbial ecology of human skin in health and disease. J Investig Dermatol Symp Proc 6(3):167–169
54. Rosselli R, Romoli O, Vitulo N, Vezzi A, Campanaro S, de Pascale F, Schiavon R, Tiarca M, Poletto F, Concheri G, Valle G, Squartini A (2016) Direct 16S rRNA-seq from bacterial communities: a PCR-independent approach to simultaneously assess microbial diversity and functional activity potential of each taxon. Sci Rep 6:32165
55. Taib N, Mangot JF, Domaizon I, Bronner G, Debroas D (2013) Phylogenetic affiliation of SSU rRNA genes generated by massively parallel sequencing: new insights into the freshwater protist diversity. PLoS One 8(3):e58950
56. Fakruddin M, Mannan K (2013) Methods for analyzing diversity of microbial communities in natural environments. Ceylon J Sci (Biol Sci) 42(1):19
57. Srinivasan R, Karaoz U, Volegova M, MacKichan J, Kato-Maeda M, Miller S, Nadarajan R, Brodie EL, Lynch SV (2015) Use of 16S rRNA gene for identification of a broad range of clinically relevant bacterial pathogens. PLoS One 10(2):e0117617
58. Alagappan A, Bergquist PL, Ferrari BC (2009) Development of a two-color fluorescence in situ hybridization technique for species-level identification of human-infectious Cryptosporidium spp. Appl Environ Microbiol 75(18):5996–5998
59. Malic S, Hill KE, Hayes A, Percival SL, Thomas DW, Williams DW (2009) Detection and identification of specific bacteria in wound biofilms using peptide nucleic acid fluorescent in situ hybridization (PNA FISH). Microbiology 155(Pt 8):2603–2611
60. Raghavachari N (2013) Microarray technology: basic methodology and application in clinical research for biomarker discovery in vascular diseases. Methods Mol Biol 1027:47–84
61. Trevino V, Falciani F, Barrera-Saldaña HA (2007) DNA microarrays: a powerful genomic tool for biomedical and clinical research. Mol Med 13(9–10):527–541
62. Miller MB, Tang YW (2009) Basic concepts of microarrays and potential applications in clinical microbiology. Clin Microbiol Rev 22(4):611–633
63. Bouzid M, Steverding D, Tyler KM (2008) Detection and surveillance of waterborne protozoan parasites. Curr Opin Biotechnol 19(3):302–306
64. Lefterova MI, Suarez CJ, Banaei N, Pinsky BA (2015) Next-generation sequencing for infectious disease diagnosis and management: a report of the association for molecular pathology. J Mol Diagn 17(6):623–634
65. Greay TL, Gofton AW, Paparini A, Ryan UM, Oskam CL, Irwin PJ (2018) Recent insights into the tick microbiome gained through next-generation sequencing. Parasit Vectors 11(1):12
66. Ghilamicael AM, Boga HI, Anami SE, Mehari T, Budambula NLM (2018) Potential human pathogenic bacteria in five hot springs in Eritrea revealed by next generation sequencing. PLoS One 13(3):e0194554
67. Moreno Y, Moreno-Mesonero L, Amorós I, Pérez R, Morillo JA, Alonso JL (2018) Multiple identification of most important waterborne protozoa in surface water used for irrigation purposes by 18S rRNA amplicon-based metagenomics. Int J Hyg Environ Health 221(1):102–111
68. Revez J, Espinosa L, Albiger B, Leitmeyer KC, Struelens MJ, ECDC National Microbiology Focal Points and Experts Group (2017) Survey on the use of whole-genome sequencing for infectious diseases surveillance: rapid expansion of European National Capacities, 2015–2016. Front Public Health 5:347
69. Aw TG, Rose JB (2012) Detection of pathogens in water: from phylochips to qPCR to pyrosequencing. Curr Opin Biotechnol 23(3):422–430
70. Ahmadian A, Ehn M, Hober S (2006) Pyrosequencing: history, biochemistry and future. Clin Chim Acta 363(1–2):83–94
71. Hong PY, Hwang C, Ling F, Andersen GL, LeChevallier MW, Liu WT (2010) Pyrosequencing analysis of bacterial biofilm communities in water meters of a drinking water distribution system. Appl Environ Microbiol 76(16):5631–5635
72. Lazcka O, Del Campo FJ, Muñoz FX (2007) Pathogen detection: a perspective of traditional methods and biosensors. Biosens Bioelectron 22(7):1205–1217
73. Connelly JT, Baeumner AJ (2012) Biosensors for the detection of waterborne pathogens. Anal Bioanal Chem 402(1):117–127

74. Bhalla N, Jolly P, Formisano N, Estrela P (2016) Introduction to biosensors. Essays Biochem 60(1):1–8
75. Vidic J, Manzano M, Chang CM, Jaffrezic-Renault N (2017) Advanced biosensors for detection of pathogens related to livestock and poultry. Vet Res 48(1):11
76. Turner AP (2013) Biosensors: sense and sensibility. Chem Soc Rev 42(8):3184–3196
77. Kumar N, Hu Y, Singh S, Mizaikoff B (2018) Emerging biosensor platforms for the assessment of water-borne pathogens. Analyst 143(2):359–373
78. Ueda H, Dong J (2014) From fluorescence polarization to Quenchbody: recent progress in fluorescent reagentless biosensors based on antibody and other binding proteins. Biochim Biophys Acta 1844(11):1951–1959
79. Martinkova P, Kostelnik A, Valek T, Pohanka M (2017) Main streams in the construction of biosensors and their applications. Int J Electrochem Sci 12:7386–7403
80. Alhadrami HA (2018) Biosensors: classifications, medical applications, and future prospective. Biotechnol Appl Biochem 65(3):497–508
81. Damborský P, Švitel J, Katrlík J (2016) Optical biosensors. Essays Biochem 60(1):91–100
82. Rathee K, Dhull V, Dhull R, Singh S (2015) Biosensors based on electrochemical lactate detection: a comprehensive review. Biochem Biophys Rep 5:35–54
83. Grieshaber D, MacKenzie R, Vörös J, Reimhult E (2008) Electrochemical biosensors—sensor principles and architectures. Sensors (Basel) 8(3):1400–1458
84. Hoß SG, Bendas G (2017) Mass-sensitive biosensor systems to determine the membrane interaction of analytes. Methods Mol Biol 1520:145–157
85. Kaittanis C, Santra S, Perez JM (2010) Emerging nanotechnology-based strategies for the identification of microbial pathogenesis. Adv Drug Deliv Rev 62(4–5):408–423
86. Mustafa F, Hassan RYA, Andreescu S (2017) Multifunctional nanotechnology-enabled sensors for rapid capture and detection of pathogens. Sensors (Basel). 17(9):E2121
87. Mody VV, Siwale R, Singh A, Mody HR (2010) Introduction to metallic nanoparticles. J Pharm Bioallied Sci 2(4):282–289
88. Billington C, Hudson JA, D'Sa E (2014) Prevention of bacterial foodborne disease using nanobiotechnology. Nanotechnol Sci Appl 7:73–83
89. Holzinger M, Le Goff A, Cosnier S (2017) Synergetic effects of combined nanomaterials for biosensing applications. Sensors (Basel). 17(5):E1010
90. Shinde SB, Fernandes CB, Patravale VB (2012) Recent trends in in-vitro nanodiagnostics for detection of pathogens. J Control Release 159(2):164–180
91. Bridle H, Balharry D, Gaiser B, Johnston H (2015) Exploitation of nanotechnology for the monitoring of waterborne pathogens: state-of-the-art and future research priorities. Environ Sci Technol 49(18):10762–10777
92. Theron J, Eugene Cloete T, de Kwaadsteniet M (2010) Current molecular and emerging nanobiotechnology approaches for the detection of microbial pathogens. Crit Rev Microbiol 36(4):318–339
93. Jain KK (2007) Applications of nanobiotechnology in clinical diagnostics. Clin Chem 53(11):2002–2009
94. Neethirajan S, Ahmed SR, Chand R, Buozis J, Nagy É (2017) Recent advances in biosensor development for foodborne virus detection. Nanotheranostics. 1(3):272–295
95. Doria G, Conde J, Veigas B, Giestas L, Almeida C, Assunção M, Rosa J, Baptista PV (2012) Noble metal nanoparticles for biosensing applications. Sensors (Basel) 12(2):1657–1687
96. Wang S, Singh AK, Senapati D, Neely A, Yu H, Ray PC (2010) Rapid colorimetric identification and targeted photothermal lysis of Salmonella bacteria by using bioconjugated oval-shaped gold nanoparticles. Chemistry 16(19):5600–5666
97. Khan SA, Singh AK, Senapati D, Fan Z, Ray PC (2011) Targeted highly sensitive detection of multi-drug resistant Salmonella DT104 using gold nanoparticles. Chem Commun (Camb) 47(33):9444–9446
98. Lee JS, Lytton-Jean AK, Hurst SJ, Mirkin CA (2007) Silver nanoparticle-oligonucleotide conjugates based on DNA with triple cyclic disulfide moieties. Nano Lett 7(7):2112–2115

99. Ness JM, Akhtar RS, Latham CB, Roth KA (2003) Combined tyramide signal amplification and quantum dots for sensitive and photostable immunofluorescence detection. J Histochem Cytochem 51(8):981–987
100. Kim GY, Son A (2010) Development and characterization of a magnetic bead-quantum dot nanoparticles based assay capable of Escherichia coli O157:H7 quantification. Anal Chim Acta 677(1):90–96
101. Ho PL (2014) Magnetic nanoparticles for pathogen detection. Pathology 46:S45
102. Chen YT, Kolhatkar AG, Zenasni O, Xu S, Lee TR (2017) Biosensing using magnetic particle detection techniques. Sensors (Basel). 17(10):E2300
103. Chung HJ, Castro CM, Im H, Lee H, Weissleder R (2013) A magneto-DNA nanoparticle system for rapid detection and phenotyping of bacteria. Nat Nanotechnol 8(5):369–375
104. Lynch M, Mosher C, Huff J, Nettikadan S, Johnson J, Henderson E (2004) Functional protein nanoarrays for biomarker profiling. Proteomics 4(6):1695–1702

Treatment of Water to Prevent Water-Associated Infectious Diseases

10

Amrita Haikerwal*, Swatantra Kumar, and Shailendra K. Saxena*

Abstract

As contaminated water is increasingly associated with infectious diseases, the basic services like safe drinking water, hygiene, and sanitation need to be monitored regularly. With the accelerating scientific advancements, detection and enumeration of water-associated pathogens are possible during emergencies and outbreaks of infectious diseases. The best recommended tests for screening of water-associated pathogens are based on detection of the nucleic acid such as PCR, 16S rRNA, FISH, and DNA microarrays. Various methods are used to reduce the microbial burden of drinking water. A number of awareness programs on basic health facilities are necessary to educate the rural population of developed and developing countries. Drugs and vaccines are available for the treatment and prevention of water-related infectious pathogens.

Keywords

Basic hygiene · Sanitation · Disinfection · Antibiotics · Public health measures

10.1 Introduction

As historically evidenced, methods and technologies used for treatment of water are not a contemporary invention. The use of straining, alum, and charcoal filtration, and boiling for improving the quality of drinking water, is a common

*Contributed equally as first author

A. Haikerwal · S. Kumar · S. K. Saxena (✉)
Centre for Advanced Research, Faculty of Medicine, King George's Medical University, Lucknow, India
e-mail: shailen@kgmcindia.edu

S. K. Saxena (ed.), *Water-Associated Infectious Diseases*,
https://doi.org/10.1007/978-981-13-9197-2_10

practice running down since 6000 years [1]. With the accelerating scientific advancements, detection and enumeration of water-associated pathogens are possible during emergencies and outbreaks of infectious diseases. However, detection of whether infectious agents belonging to the same family and the strains are pathogenic or nonpathogenic is challenging. The strategies required for proper analysis and assessment of microbial population in water should be robust sensitive and with high throughput. The novel interventions for detecting infectious pathogens to maintain the standard of drinking water are imperative.

The best recommended tests for the identification of water-associated pathogens are based on detection of the nucleic acid such as PCR, 16S rRNA, FISH, and DNA microarrays. PCR is one of the most frequently used techniques for the detection of water-associated pathogens [2]. 16S rRNA is based on the detection of bacterial species based on genetic library [3]. Fluorescence in situ hybridization (FISH) is based on the principle of hybridization of rRNA probes tagged with fluorescent dye [4]. The most recent molecular technique is DNA microarrays which can be used to test diverse water samples simultaneously [5]. Although molecular techniques are efficient methods to detect microbial pathogens all of them are ineffective in detection of viral pathogens and toxins including extracellular products of pathogens existing in water.

Newer techniques are required which combine the strategy of accuracy assigned with traditional microbiology and sensitivity related with molecular biology. Biosensor technology has provided a platform which brings together both sensitivity and accuracy with rapid detection. It also provides the opportunity of real-time monitoring of infectious agents present in the water bodies which allows the preventive measure to be compelled prior to any threat in future. Label-free biosensors and peptide-based receptor molecule probes are the most effective strategies which are based on the various technologies of sensor platform [6]. Similarly, the early diagnosis of infectious diseases is the key to fight against it; thus, use of nano-diagnostics has proven to be promising in the detection of infectious pathogens as compared to conventional diagnostic tools. Nano-diagnostic methods are sensitive and cost effective that might be provided in developing countries deprived of resources to fund high-cost molecular devices or a vaccination program for the protection of millions of people or that are dependent on international support [7]. However, nano-diagnostics could be used in remote areas as point-of-care unit for the diagnosis of infectious diseases. Numerous studies have reported usage of nanoparticle base immunosensor for the rapid screening of infectious pathogens; for instance, *Vibrio cholerae* was detected by magnetite nanoparticles capped with citric acid, deposited onto glass substrate coated with indium tin oxide, and immobilized with monoclonal antibodies against *Vibrio cholerae* [8].

10.2 Steps Towards Preventing Water-Associated Infectious Diseases

In the growing number of cases of water-borne infectious diseases, there is requirement of monitoring basic services of safe drinking water and sanitation across the community.

10.2.1 Improved Water Sources

A regular cleaning, maintenance, and protection of piped water as well as non-piped water via hand-dug wells, rain water, boreholes, tankers, and packaged and bottled water are essential. Rivers, lakes, and streams that replenish the water table should be implicated. In addition, water should be stored in high-capacity tanks or are supplied directly to houses. Pollution by ripple dust, clothing washing, solid waste infiltration, and industrial and animal bathing chemical products, such as buffalos, deteriorate the quality of water [9].

10.2.2 Improving the Water at the Point of Use

Drinking water free from all types of contamination is the fundamental requirement without any inequality. Pretreatment of water, adulterated due to handling and storage problems, can also be used for preparation of food or drinking purpose [10]. Numerous methods for minimizing the microbial count of drinking water are used in combinations.

- Aeration (vigorously shaking) increases the O_2 content and helps in evacuation of H_2S and CH_4 that maintains the odor and taste of water [11].
- Storage and sedimentation of water bring down the suspended solid particles and reduce the antigenicity of the pathogens.
- Filtration comprising membrane filters for removal of pathogens as small as viruses [12], sand filters, and ceramic filters [13] which are generally loaded with silver (antibacterial) blocks the physical entities existing in water.
- Disinfection of water is carried out by boiling and using chemicals such as chlorine (according to the standard formulation) for removing infectious water-associated pathogens [14].
- UV rays (sunlight) for appropriate exposure time and intensity are used for killing microbes in water [15] (Fig. 10.1).

10.2.3 Maintenance of Basic Hygiene Facility

The basic hygiene facilities comprise proper hand-washing with soap and water and maintenance of menstrual and food hygiene. In order to educate people, especially in rural areas of developing countries, different awareness programs concerning basic hygiene facilities are required. The establishment of hand-washing practice after defecation is the priority area of global agencies [16].

10.2.4 Limiting the Vector Transmission of Water-Associated Infectious Pathogens

There is an alarming rate in the global burden of mosquito-transmitted infectious diseases. Stagnant contaminated water, which is stored in utensils, is the residence

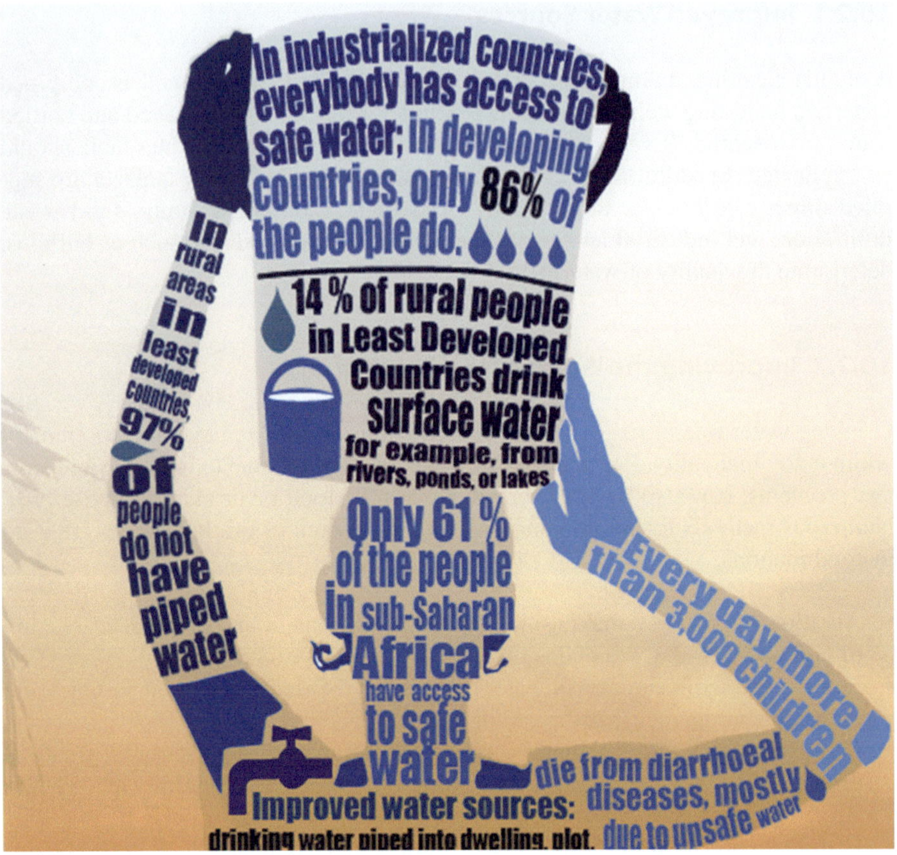

Fig. 10.1 The data representing the usage of unsafe water in developing countries [adapted from WHO]

of mosquito eggs, larvae, and other products. Several water-borne infectious diseases like Japanese encephalitis, dengue, and malaria are spread by mosquito vectors for completion of pathogen's vicious life cycle (Fig. 10.2). The control program of mosquito vector includes several approaches: avoidance of water logging, repellents, and impregnated nets treated with insecticides, larvivores, and *Gambusia affinis*, and disinfection of their breeding habitats [17].

10.2.5 Disposal of Excreta and Management of Solid Waste and Dead Bodies

Toilet use will stop the open defecation and prevent fecal of water contamination in each household and public place [18]. Likewise, the behavior of society regarding cremating dead bodies in river and burying them in ground needs to be changed and strict prohibition will maintain the quality of water. Moreover, management of heaps of solid wastes will prevent germination of mosquitoes and fungus resulting in prevention of respiratory diseases [19].

Fig. 10.2 Vector prevention methods [adapted from WHO]

10.2.6 Drugs and Vaccines

The most robust and promising methods are to treat and prevent water-related infectious pathogens with drugs and vaccines. Some of the vaccines available are, for instance, malaria vaccine: Mosquirix or also known as RTS, S/AS01 (RTS, S) [20]; typhoid vaccine: inactivated vaccine given as a shot; live attenuated vaccine given orally [21, 22]; and dengue vaccine: Dengvaxia, DENVax, LATV Δ30 [23, 24]. The drugs approved for the treatment of these diseases include praziquantel for schistosomiasis [25, 26], miltefosine for amoebic meningoencephalitis [27, 28], suramin for dengue, and 5-nitroimidazolic for giardiasis [29].

References

1. World Health Organization (2003) Emerging issues in water and infectious disease. WHO, Geneva. www.who.int/water_sanitation_health/emerging/emerging.pdf. Accessed on 6-11-2017
2. Lee SH, Joung M, Yoon S, Choi K, Park WY, Yu JR (2010) Multiplex PCR detection of water-borne intestinal protozoa: Microsporidia, Cyclospora, and Cryptosporidium. Korean J Parasitol 48(4):297–301
3. Dick LK, Field KG (2004) Rapid estimation of numbers of fecal Bacteroidetes by use of a quantitative PCR assay for 16S rRNA genes. Appl Environ Microbiol 70(9):5695–5697
4. Moreno Y, Ferrús MA, Alonso JL, Jiménez A, Hernández J (2003) Use of fluorescent in situ hybridization to evidence the presence of Helicobacter pylori in water. Water Res 37(9):2251–2256
5. Lee DY, Seto P, Korczak R (2010) DNA microarray-based detection and identification of water-borne protozoan pathogens. J Microbiol Methods 80(2):129–133
6. Dover JE, Hwang GM, Mullen EH, Prorok BC, Suh SJ (2009) Recent advances in peptide probe-based biosensors for detection of infectious agents. J Microbiol Methods 78(1):10–19
7. Coloma J, Harris E (2009) Molecular genomic approaches to infectious diseases in resource-limited settings. PLoS Med 6(10):1–6
8. Sharma A, Baral D, Rawat K, Solanki PR, Bohidar HB (2015) Biocompatible capped iron oxide nanoparticles for Vibrio cholerae detection. Nanotechnology 26(17):175302
9. World Health Organization (2013) Cleaning and disinfecting water storage tanks and tankers. WHO, Geneva. http://www.who.int/water_sanitation_health/publications/2011/tn3_cleaning_disinfecting_tanks_en.pdf. Accessed on 7-11-2017
10. World Health Organization (2013) Emergency treatment of drinking-water at the point of use. WHO, Geneva. http://www.who.int/water_sanitation_health/publications/2011/tn5_treatment_water_en.pdf. Accessed on 7-11-2017
11. Daelman MR, van Voorthuizen EM, van Dongen UG, Volcke EI, van Loosdrecht MC (2012) Methane emission during municipal wastewater treatment. Water Res 46(11):3657–3670
12. Metreveli G, Wågberg L, Emmoth E, Belák S, Strømme M, Mihranyan A (2014) A size-exclusion nanocellulose filter paper for virus removal. Adv Healthc Mater 3(10):1546–1550
13. van der Laan H, van Halem D, Smeets PW, Soppe AI, Kroesbergen J, Wubbels G, Nederstigt J, Gensburger I, Heijman SG (2014) Bacteria and virus removal effectiveness of ceramic pot filters with different silver applications in a long term experiment. Water Res 51:47–54
14. Fiorentino A, Ferro G, Alferez MC, Polo-López MI, Fernández-Ibañez P, Rizzo L (2015) Inactivation and regrowth of multidrug resistant bacteria in urban waste water after disinfection by solar-driven and chlorination processes. J Photochem Photobiol B 148:43–50
15. Nguyen MT, Jasper JT, Boehm AB, Nelson KL (2015) Sunlight inactivation of fecal indicator bacteria in open-water unit process treatment wetlands: modeling endogenous and exogenous inactivation rates. Water Res 83:282–292
16. World Health Organization (2013) Hygiene promotion in emergencies. WHO, Geneva. http://www.who.int/water_sanitation_health/emergencies/WHO_TN_10_Hygiene_promotion_in_emergencies.pdf. Accessed on 08-3-2018
17. Haikerwal A, Bhatt MLB, Saxena SK (2017) Reducing the global burden of dengue: steps toward preventive methods. Arch Prev Med 2(1):028–033

18. World Health Organization (2013) Technical options for excreta disposal in emergencies. http://www.who.int/water_sanitation_health/emergencies/WHO_TN_14_Technical_options_for_excreta_disposal.pdf. Accessed on 08-3-2018
19. World Health Organization (2013) Disposal of dead bodies in emergency conditions. http://www.who.int/water_sanitation_health/publications/tech_note8/en/index2.html. Accessed on 12-3-2018
20. The RTS,S malaria vaccine candidate (2017) PATH. http://www.malariavaccine.org/sites/www.malariavaccine.org/files/content/page/files/mviCVIA_rtss.pdf. Accessed on 12-3-2018
21. Typhoid vaccines (2012) CDC. https://www.cdc.gov/vaccines/hcp/vis/vis-statements/typhoid.pdf. Accessed on 8-3-2018
22. World Health Organization (2003) Background document: the diagnosis, treatment and prevention of typhoid fever. http://www.who.int/rpc/TFGuideWHO.pdf. Accessed on 8-3-2018
23. Saxena SK, Haikerwal A, Gadugu S, Bhatt MLB (2017) Complementary and alternative medicine in alliance with conventional medicine for dengue therapeutics and prevention. Future Virol 121(8):399–402
24. Haikerwal A, Kumar S, Kant R, Saxena SK (2017) Potential therapeutics for dengue virus infection. Ann Pharmacol Pharm 2(8):1–6
25. World Health Organization (2017) Schistosomiasis. WHO, Geneva. http://www.who.int/mediacentre/factsheets/fs115/en/. Accessed on 8-3-2018
26. Barda B, Coulibaly JT, Hatz C, Keiser J (2017) Ultrasonographic evaluation of urinary tract morbidity in school-aged and preschool-aged children infected with Schistosoma haematobium and its evolution after praziquantel treatment: a randomized controlled trial. PLoS Negl Trop Dis 11(2):1–13
27. Cope JR, Conrad DA, Cohen N, Cotilla M, DaSilva A, Jackson J, Visvesvara GS (2016) Use of the novel therapeutic agent Miltefosine for the treatment of primary Amebic meningoencephalitis: report of 1 fatal and 1 surviving case. Clin Infect Dis 62(6):774–776
28. Linam WM, Ahmed M, Cope JR, Chu C, Visvesvara GS, da Silva AJ, Qvarnstrom Y, Green J (2015) Successful treatment of an adolescent with Naegleria fowleri primary amebic meningoencephalitis. Pediatrics 135(3):2014–2292
29. Coronato-Nunes B, Calegar DA, Monteiro KJ, Hubert-Jaeger L, Reis ER, das Chagas Xavier SC, Carpp LN, Lima MM, Bóia MN, Carvalho-Costa FA (2017) Giardia intestinalis infection associated with malnutrition in children living in northeastern Brazil. J Infect Dev Ctries 11(7):563–570

Conventional Treatments of Water-Associated Infectious Diseases

11

Vimal K. Maurya*, Swatantra Kumar, and Shailendra K. Saxena*

Abstract

Water-borne infectious diseases are the major threats to the population living in the developing countries. Treatment of water-associated infectious diseases mainly involves patient care and supportive therapy which includes the symptomatic management of the diseases. Numerous drugs are undergoing clinical trials to evaluate their efficacy against several water-associated infections but still there is an unmet gap. For bacterial and parasitic infections, antibiotics are given and found to be effective against these infections whereas in most cases of viral infections no licensed antiviral drug is available. But the development of multiple drug resistance in water-associated pathogens has posed new challenge for the complete management of these infections. World Health Organization (WHO) and Centers for Disease Control and Prevention (CDC) have recommended various guidelines for the treatment and prevention of water-associated infectious pathogens using drugs and vaccines which are the most robust and promising way to restrict these infections. Some of the vaccines available are against water-associated infectious diseases such as malaria vaccine: Mosquirix or also known as RTS, S/AS01 (RTS, S); typhoid vaccine: inactivated vaccine given as a shot; live attenuated vaccine given orally; oral cholera vaccines (OCV); oral polio vaccine; and dengue vaccine: Dengvaxia, DENVax. The drugs approved for the treatment of these diseases include doxycycline for cholera, miltefosine for amoebic meningoencephalitis, azithromycin for *Campylobacter jejuni*, and nitazoxanide for cryptosporidiosis.

*Contributed equally as first author

V. K. Maurya · S. Kumar · S. K. Saxena (✉)
Centre for Advanced Research, Faculty of Medicine, King George's Medical University, Lucknow, India
e-mail: shailen@kgmcindia.edu

S. K. Saxena (ed.), *Water-Associated Infectious Diseases*,
https://doi.org/10.1007/978-981-13-9197-2_11

Keywords
Norfloxacin · Ciprofloxacin · Trimethoprim · β-Lactam antibiotics · Ceftriaxone Doxycycline · Clarithromycin · Amoxicillin · Antispasmodic drugs

11.1 Introduction

11.1.1 Typhoid

Antibiotics such as cephalosporins, fluoroquinolones, and azithromycin are effective in the treatment of typhoid fever [1]. Vaccines are available to control endemic typhoid fever and future outbreak. Typhoid conjugate vaccines (TCV) consisting of VI polysaccharide antigen associated with tetanus toxoid protein are suitable for all age-group patients due to its improved immunological characteristics and longer duration of protection [2]. Apart from typhoid vaccination program, other efforts such as improvement of water quality, health education, and training of health professionals in diagnosis and treatment are important to control the disease [3].

11.1.2 Bacillus Dysentery

Antibiotics such as ciprofloxacin and azithromycin are beneficial in severe cases of shigellosis. Generally, most of the people recover from shigellosis without treatment in 5–7 days. Maintenance of water and electrolyte balance and treatment with bismuth subsalicylate help to reduce the severity of disease in chronic cases of shigellosis [4]. There is no candidate vaccine which is found to be effective against Shigella. Most of the vaccines made against *S. flexneri* are under clinical investigation. Measles immunization can extensively minimize the prevalence and severity of diarrheal diseases [5].

11.1.3 Cholera

Cholera is an easily curable disease. Management of cholera is based on rehydration therapy, antibiotic treatment, and zinc supplementation. Patients with cholera infection can be treated effectively through administration of oral rehydration solution (ORS). The WHO/UNICEF ORS standard sachet is dissolved in 1 liter (L) of clean water and given to the patient. Antibiotic treatment is needed for severe cases of cholera. Zinc supplementation is an important adjunctive therapy for children, which reduces the duration and may prevent future incident of infection [6]. Currently, three WHO pre-qualified oral cholera vaccines (OCV) are approved for the prevention of cholera: Euvichol®, Shanchol™, and Dukoral®. All the three vaccines needed two doses for complete protection against cholera infection [7]. A recommendation for the use of antibiotics for the treatment of cholera has been mentioned in Table 11.1.

Table 11.1 Recommended antibiotics for the treatment of cholera [8–11]

Organization	Recommendation	First-line drug	Alternate drug	Drug choices
World Health Organization	Antibiotic treatment for cholera patients with severe dehydration only	Doxycycline	Tetracycline	Erythromycin is recommended drug for children
Pan American Health Organization	Antibiotic treatment for cholera patients with moderate or severe dehydration	Doxycycline	CiprofloxacinAzithromycin	Erythromycin or azithromycin recommended as first-line drugs for pregnant women and children. Ciprofloxacin and doxycycline recommended as second-line drugs for children
International Centre for Diarrhoeal Disease Research, Bangladesh	Antibiotic treatment for cholera patients with some or severe dehydration	Doxycycline	CiprofloxacinAzithromycinCo-trimoxazole	Erythromycin recommended as first-line drug for children and pregnant women
Medecins Sans Frontieres	Antibiotic treatment for severely dehydrated patients only	Doxycycline	ErythromycinCo-trimoxazoleChloramphenicol	

11.1.4 Enteropathogenic *E. coli*

Maintenance of water and electrolyte balance and treatment with antibiotics are required during *Enteropathogenic E. coli* infection. Patients having profuse diarrhea or vomiting should be rehydrated and given intravenous fluids which reduces the risk of oligoanuric renal failure. Antibiotics such as fluoroquinolones (ciprofloxacin) and macrolides (azithromycin, and rifaximin) are used to treat *E. coli* infection. Antimotility agents are not recommended in case of bloody diarrhea and STEC infection [12].

11.1.5 Aeromonas

Infection can be treated with antibiotics. All strains of Aeromonas show resistance against ampicillin but they are susceptible to chloramphenicol, tetracycline, nalidixic acid, cefotaxime, trimethoprim-sulfamethoxazole, and ciprofloxacin treatment. Patients suffering from persistent diarrhea should be treated with norfloxacin, ciprofloxacin, and trimethoprim-sulfamethoxazole. Temperatures below 14 °C, free chlorine residuals above 0.1–0.2 mg/L, and restriction of organic compounds that act as nutrients are the key factors in controlling Aeromonas proliferation [13].

11.1.6 Legionellosis

Legionellosis is a nonpneumonic form of infection and treated with antibiotics including macrolides, tetracyclines, ketolides, and quinolones. Choice of antibiotics depends upon diagnosis, and severity of infection. Legionnaires' disease can be prevented by applying control measures to decrease the growth of Legionella and dissemination of aerosols. There is no vaccine currently available for Legionnaires' disease [14].

11.1.7 Coxsackie Virus A

No accepted vaccine or antiviral drug is available for Coxsackie A virus. Cure is mainly based on symptoms arising during infection. For management of pain and fever paracetamol is given. Nonsteroidal anti-inflammatory drugs such as aspirin or ibuprofen are contraindicated during Coxsackie A virus infection [15]. Use mouthwashes or sprays helpful in reducing mouth sores. People with Coxsackie virus disease should maintain fluid and electrolyte balance to prevent dehydration.

11.1.8 Coxsackie Virus B

There is no licensed antiviral agent or vaccine available for the management of Coxsackie B virus. Supportive treatment is available according to the severity of infection [16]. Symptomatic cure may reduce the severity of infection. For the management of aseptic meningitis drugs such as pleconaril are found to be effective [17]. Carvedilol is a nonselective beta-blocker given for the treatment of myopericarditis and has shown to inhibit the replication of the virus in a murine model but this has not been tested in humans. Analgesics and narcotics have been shown to be helpful in patients suffering from epidemic pleurodynia [18].

11.1.9 Echoviruses

There is no specific antiviral for viral meningitis; the available treatment options are mostly supportive. Rest, hydration, antipyretics, and analgesics or anti-inflammatory medications may be given as needed. There is no vaccine to protect against echoviruses that are the foremost cause of viral meningitis in humans. Vaccination against other diseases which are also known to cause viral meningitis such as influenza, measles, chickenpox, and mumps may provide protection against echoviruses [19].

11.1.10 *Yersinia enterocolitica*

Most infections are self-limited. Antibiotics should be provided in severe cases. *Y. enterocolitica* is usually perceptive to fluoroquinolones, tetracyclines, trimethoprim-sulfamethoxazole, third-generation cephalosporins, and aminoglycosides [20]. *Y. enterocolitica* is usually resistant to first-generation cephalosporins and most penicillin. Antimicrobial therapy is not effective on postinfectious sequelae [21].

11.1.11 *Campylobacter jejuni*

Campylobacter infection is generally a mild, self-limited infection. Maintenance of fluid and electrolyte balance is the main focus during the treatment of infection [22]. In most cases of *C. jejuni* gastroenteritis antibiotics are not required while in severe cases azithromycin therapy would be a primary antibiotic choice for Campylobacter infections [23].

11.1.12 *Pseudomonas aeruginosa*

Antibiotics with potent antipseudomonal activity are available. Drugs such as aminoglycosides, ceftazidime, carbapenems, ureidopenicillins, cefepime, aztreonam, and ciprofloxacin are frequently used to treat *P. aeruginosa* infection. While treating severe pseudomonal infection with β-lactams, maintenance optimal dose and duration of action of drug are important [24].

11.1.13 Leptospirosis

Antimicrobial therapy is effective against leptospirosis and should be initiated early to reduce the severity and duration of infection. β-Lactam antibiotics such as intravenous penicillin and ceftriaxone are used for the treatment of severe leptospirosis. In severe cases of leptospirosis hospitalization and supportive therapy are required which includes fluid and electrolyte supplementation via intravenous hydration [25]. For patients with oliguric renal failure and respiratory failure, dialysis or mechanical ventilation is provided. Doxycycline is a drug of choice for patients with mild symptoms, but this drug is contraindicated in pregnant women and children. No vaccine is available and the paramount way to reduce transmission of infection is to avoid exposure like contact with potentially contaminated bodies of water, walking in flood waters, and contact with potentially infected animals or their body fluids [26].

11.1.14 *Helicobacter pylori*

Patients with active duodenal or gastric ulcers caused by *H. pylori* should be treated with antibiotics. Treatment should be specified on an individual and based on severity of infection. Standard treatments are clarithromycin triple therapy (proton pump inhibitor [PPI]) + amoxicillin or metronidazole or clarithromycin + bismuth quadruple therapy (PPI or H2-blocker + bismuth + metronidazole + tetracycline) [27, 28].

11.1.15 Polio Virus

There is no licensed antiviral drug for polio virus. Treatment is mainly based on symptoms arising during infection [29]. Antispasmodic drugs are given for treatment of muscular pain. Vaccines are available and given multiple times to protect the child from polio infection. There are six different vaccines which are shown to be protective against polio spread: trivalent oral polio vaccine (tOPV) and inactivated polio vaccine (IPV) protecting against poliovirus types 1, 2, and 3 whereas bivalent oral polio vaccine (bOPV) protecting against poliovirus types 1 and 3. Similarly, monovalent oral polio vaccines (mOPV1, mOPV2, and mOPV3) protect against each individual type of polio virus, respectively. To prevent the transmission and future outbreaks of polio viruses, high levels of the vaccination should be maintained [30].

11.1.16 Rotaviruses

No specific antiviral drug yet has been licensed for the rotavirus infection. People with rotavirus infection should increase fluid intake to prevent dehydration [31]. Rotavirus vaccine is found to be protective against rotavirus infection. The vaccine will be protected against serious rotavirus infection in most children. RotaTeq® (RV5) and Rotarix® (RV1) are the two rotavirus vaccines currently licensed for infants in the USA. Rotarix® (RV1) is given in two doses at ages 2 and 4 months and RotaTeq® (RV5) is given in three doses at ages 2, 4, and 6 months [32].

11.1.17 Adenoviruses

There is no specific treatment for adenoviruses. Sympathetic treatment is given in mild cases of adenovirus infections. In severe and immunocompromised patients Cidofovir is helpful for the management of adenovirus infection. There is no vaccine approved for adenoviruses [33].

11.1.18 Hepatitis A

There is no licensed antiviral for hepatitis A virus infection and only supportive therapy is available for the management of infection [34]. Treatment may take several weeks or months for complete cure of infection. Maintaining comfort and adequate nutritional balance of the patient reduces the severity of infection. During infection fluid and electrolyte balance should be maintained that are lost from vomiting and diarrhea. A live oral vaccine is also available in China but its efficacy against hepatitis A virus infection is still under investigation. There are various inactivated hepatitis A vaccines available and all are similar in terms of protection of people from the virus and their side effects [35].

11.1.19 Hepatitis E

There is no specific antiviral licensed for hepatitis E virus infection. The disease is normally self-constraining and hospitalization is not required in acute infections but in severe cases of hepatitis E hospitalization is required. Ribavirin is the drug of choice for a patient who is immunosuppressed and has chronic hepatitis E. In some cases, interferon has shown to be effective against hepatitis E infection. A recombinant subunit vaccine against hepatitis E virus infection was registered in China in 2011 [36].

11.1.20 Norovirus

There is no selective antiviral drug for norovirus and management is only based on supportive care, especially oral or intravenous rehydration. In case of severe vomiting, antiemetic agents should be provided to the patients. Antimotility agents should be contraindicated in infants but may be useful in older children and adults. No vaccine is currently available for the prevention of norovirus. The most efficient way of reducing norovirus infection is to wash hands with soap and water for at least 20 s. Although alcohol-based hand sanitizers might be helpful between handwashings they should not be considered an alternative for soap and water [37].

11.1.21 *Cryptosporidium hominis*

Most of the patients will recover without treatment. Nitazoxanide is licensed for cryptosporidiosis treatment in patients under the age of 1 year [38]. Nitazoxanide has shown to be less effective against cryptosporidiosis in patients with weak immune response. Protease inhibitors have shown direct anti-Cryptosporidium activity. Oral rehydration is the most effective supportive therapy in both immune-competent and immune-compromised patients. Oral rehydration should be provided in case of diarrhea and antidiarrheal medication should be given according to the severity of infection [39].

11.1.22 *Entamoeba histolytica*

Metronidazole or tinidazole is used for symptomatic intestinal and extraintestinal infection. In case of drug resistance disease should be treated with iodoquinol or paromomycin. Asymptomatic patients infected with *E. histolytica* should be treated with iodoquinol or paromomycin which also reduces the risk of disease transmission. Patients with *E. histolytica* should be rehydrated and given intravenous fluids which decrease the risk of oligoanuric renal failure [40].

11.1.23 *Giardia lamblia*

The antimicrobial drugs such as metronidazole, tinidazole, and nitazoxanide are effective against *G. lamblia* infection. In case of drug resistance alternative medications like paromomycin, quinacrine, and furazolidone are used to treat *G. lamblia* infection. Selection of drugs is based on different factors such as medical history, nutritional status, and condition of the immune system [41].

11.1.24 *Naegleria fowleri*

Combinations of drugs with different regimens are used for the management of *N. fowleri* infection: Miltefosine is recommended by CDC in combination with other drugs such as trimethoprim/sulfamethoxazole (antibiotic), sulfadiazine (antibiotic), pentamidine (antifungal and antiprotozoal drug), flucytosine (antifungal drug), fluconazole or the related drugs voriconazole or itraconazole (antifungal drugs), and amphotericin B (antifungal drug). CDC also advises that miltefosine and amphotericin B deoxycholate are the backbone of treatment regimen for this infection [42, 43].

11.1.25 *Ascaris lumbricoides*

Anthelminthic medications such as mebendazole and albendazole are the drugs of choice for treatment of Ascaris infections. For the complete eradication, infections are generally treated for 1–3 days with anthelmintic drugs. Levamisole and pyrantel are the two most effective alternatives in case of drug resistance. Pyrantel can be given in patients of all ages. Ivermectin is also given but its safety for treating children has not been established [44].

11.1.26 Malaria

Malaria caused by the *P. falciparum* is treated with artemisinin-based combination therapies (ACTs). The selection of ACT is based on therapeutic efficacy against *P. falciparum*. In antimalarial regimens, a single low dose of primaquine should be added in order to decrease the infection spread in low transmission areas. Similarly, malaria caused by *P. vivax* should be treated with an ACT or chloroquine whereas in regions where chloroquine-resistant *P. vivax* has been recognized infections should be treated with ACT. All ACTs are effective against blood-stage infections of *P. vivax* except artesunate+sulfadoxine-pyrimethamine (AS+SP) combination. In recent years, admittance to ACTs has been increased significantly. 80 countries had adopted ACTs as first-line treatment by the end of 2016 [45, 46]. Mosquirix™ also known as RTS, S/AS01, was licensed by the European Medicines Agency for active immunization in children against malaria.

11.1.27 Filaria

WHO recommended mass drug administration (MDA) for the eradication of lymphatic filariasis which involves giving an annual dose of antifilarial drugs to the whole risk population. Although antifilarial drugs have a limited effect on adult parasites they sufficiently decrease the density of microfilariae in the bloodstream and also restrict the transmission of parasites to mosquitoes. The recommendation of MDA regimen depends on the co-endemicity of lymphatic filariasis with other filarial diseases. WHO recommends the following MDA regimens: albendazole (400 mg) alone twice per year for areas co-endemic with loiasis, ivermectin (200 mcg/kg) with albendazole (400 mg) in countries with onchocerciasis, and diethylcarbamazine citrate (DEC) (6 mg/kg) and albendazole (400 mg) in countries without onchocerciasis. Diethylcarbamazine is another drug of choice for the treatment of filaria [47, 48].

11.1.28 Japanese Encephalitis

No licensed antiviral agent is available for JE infection. The treatment depends on supportive care and managing the neurological complications associated with JE [49]. Vaccines are available and found to be safe and effective against JE infection. Currently, four types of JE vaccines are in use: live recombinant vaccines, live attenuated vaccines, inactivated Vero cell-derived vaccines, and inactivated mouse brain-derived vaccines. The SA14-14-2 live attenuated vaccine developed in China has been the most commonly used in endemic countries in recent years [50].

11.1.29 Dengue Fever

No specific antiviral agents are licensed for dengue fever. During the infection, fever is controlled with acetaminophen and tepid sponge baths. NSAIDs like aspirin and ibuprofen are contraindicated during infection because they can increase the risk of bleeding [51]. Dengvaxia (CYD-TDV) is a live recombinant tetravalent dengue vaccine developed by Sanofi Pasteur and this vaccine has been licensed for use in persons between 9 and 45 years of age living in endemic areas such as Mexico, the Philippines, Brazil, and Thailand. Personal protective measures such as wearing clothing on arms and legs can prevent dengue infection in the early morning and late afternoon in particular if the risk is greatest to be bitten [52].

11.1.30 Zika Virus

There is no particular antiviral agent or vaccine available for Zika virus infection. Disease is usually mild and requires no specific treatment. Patients infected with this virus should take adequate rest and maintain the fluid and electrolyte balance. For the management of fever paracetamol is given to the patients. Aspirin and other NSAIDs should be avoided. As there is no vaccine available, it is important to avoid mosquito exposure by using personal protective measures and simultaneously educating the peoples about the risk of Zika virus infection [53].

11.1.31 West Nile Virus

There is no specific therapy to reduce the injury or resulting edema associated with WNV infection. The clinical management is only supportive care. Some studies reported that ribavirin and interferon-α 2b have demonstrated efficacy against the WNV in cell culture model but there are no clinical trials to support the efficacy of either agent [54]. Patients with severe meningeal complications should be treated with paracetamol and antiemetic drugs. In case of nausea and vomiting, adequate rehydration and electrolyte balance are beneficial. As no WNV vaccine is licensed for human use, the most efficient way of reducing infection is to wash hands with soap and water. Although alcohol-based hand sanitizers might be helpful between hand-washings they should not be considered an alternative for soap and water. The personal protective measures include effective mosquito repellents (those containing DEET), avoiding mosquito biting, applying barrier methods such as long-sleeve clothing and long pants, and screening of blood and organ donors to decrease exposure to infected mosquitoes [55].

11.1.32 Yellow Fever Infection

There is no licensed antiviral drug available for yellow fever and management of infection only depends on supportive therapy which includes the treatment of dehydration, liver and kidney failure, and fever improving the outcomes. Antibiotics are used for the treatment of other bacterial infections associated with yellow fever infection. Through proper vaccination yellow fever can be prevented [56]. 17D is a live attenuated viral vaccine which is a highly effective and only commercially available vaccine for yellow fever infection. In infants less than 6 months of age, the yellow fever vaccine is contraindicated except when the risk of yellow infection is very high [57].

11.2 Conclusions and Future Perspectives

To decrease the burden of water-associated infectious disease various precautionary measures have to be adapted for the sustainable solution. Implementation of guidelines for prevention will reduce mortality and disability due to nosocomial infections, occasional infections, resistance to antimicrobial disease, and newborn infections. Similarly, new pathogens associated with re-emerging and drug-resistant water will develop improved diagnostic testing methods. New approaches will be developed to reduce the populations of insect vectors and control animal stocks for human diseases. Sometimes people choose their own medications by consulting a doctor which may be risky as the duration of the disease increases. A better understanding of the risk factors in developing infections and diseases will offer new opportunities to prevent diseases and to understand the relationships between infectious agents and certain chronic diseases.

References

1. Centers for Disease Control and Prevention (CDC) (2017) Typhoid and paratyphoid fever, 31 May 2017. https://wwwnc.cdc.gov/travel/yellowbook/2018/infectious-diseases-related-totravel/typhoid-paratyphoid-fever. Accessed on 12-10-2018
2. World Health Organization (2018) Immunization, vaccines and biologicals for Typhoid fever, 11 Sept 2018. http://www.who.int/immunization/diseases/typhoid/en/. Accessed on 12-10-2018
3. World Health Organization (2018) Media centre for Typhoid fever, Jan 2018. http://www.who.int/mediacentre/factsheets/typhoid/en/. Accessed on 12-10-2018

4. World Health Organization (2005) Guidelines for the control of shigellosis, including epidemics due to Shigella dysenteriae type 1. http://apps.who.int/iris/bitstream/handle/10665/43252/9241592330.pdf;jsessionid=B90A37B7E7B0812CE876C762E438CEBA?sequence=1. Accessed on 12-10-2018
5. Centers for Disease Control and Prevention (2018) Diagnosis and treatment of Shigellosis, 7 Jun 2018. https://www.cdc.gov/shigella/diagnosistreatment.html. Accessed on 12-10-2018
6. Centers for Disease Control and Prevention (CDC) (2015) Treatment of cholera, 20 Jan 2015. https://www.cdc.gov/cholera/treatment/antibiotic-treatment.html. Accessed on 13-10-2018
7. World Health Organization (2018) Oral cholera vaccines for the prevention of cholera, 1 Feb 2018. http://www.who.int/news-room/fact-sheets/detail/cholera. Accessed on 13-10-2018
8. WHO (2004) Cholera outbreak: assessing the outbreak response and improving preparedness. http://www.who.int/cholera/publications/final%20outbreak%20booklet%20260105-OMS.pdf. Accessed on 13-10-2018
9. PAHO (2010) Recommendations for clinical management of cholera, 4 Nov 2010. www.paho.org/hq/dmdocuments/2011/cholera-clin-management-ENG-rev-JUN-1.pdf. Accessed on 13-10-2018
10. Cholera Outbreak Training and Shigellosis (COTS) program (2018). http://www.cotsprogram.com. Accessed on 13-10-2018
11. Médecins Sans Frontières (2004) Cholera guidelines, 2nd edn. http://www.bvsde.paho.org/texcom/cd045364/choleraguide.pdf. Accessed on 13-10-2018
12. Centers for Disease Control and Prevention (CDC) (2017) Treatment of Enteropathogenic E. coli, 31 May 2017. https://wwwnc.cdc.gov/travel/yellowbook/2018/infectious-diseases-related-to-travel/escherichia-coli-diarrheagenic. Accessed on 13-10-2018
13. Vila J, Ruiz J, Gallardo F, Vargas M, Soler L, Figueras MJ et al (2003) Aeromonas spp. and Traveler's diarrhea: clinical features and antimicrobial resistance. Emerg Infect Dis 9(5):552–555. https://doi.org/10.3201/eid0905.020451
14. Centers for Disease Control and Prevention (CDC) (2017) Leishmaniasis, cutaneous Legionellosis (Legionnaires' Disease & Pontiac Fever), 31 May 2017. https://wwwnc.cdc.gov/travel/yellowbook/2018/infectious-diseases-related-to-travel/legionellosis-legionnaires-disease-pontiac-fever. Accessed on 14-10-2018
15. Centers for Disease Control and Prevention (CDC) (2017) Hand foot, & mouth disease, 12 Jun 2017. https://wwwnc.cdc.gov/travel/yellowbook/2018/infectious-diseases-related-to-travel/hand-foot-and-mouth-disease. Accessed on 14-10-2018
16. Centers for Disease Control and Prevention (CDC) (2019) Hand, foot, and mouth disease: prevention & treatment. www.cdc.gov/hand-foot-mouth/about/prevention-treatment.html. Accessed on 22-07-2019
17. Irani DN (2008) Aseptic meningitis and viral myelitis. Neurol Clin 26(3):635–655, vii–viii. https://doi.org/10.1016/j.ncl.2008.03.003
18. Sagar S, Liu PP, Cooper LT Jr (2012) Myocarditis. Lancet 379(9817):738–747. https://doi.org/10.1016/S0140-6736(11)60648-X
19. Centers for Disease Control and Prevention (CDC) (2018) Viral Meningitis: prevention and treatment, 31 Jul 2018. https://www.cdc.gov/meningitis/viral.html. Accessed on 15-10-2018
20. Centers for Disease Control and Prevention (CDC) (2016) Yersinia enterocolitica (Yersiniosis): information for health & lab professionals, 25 May 2016. https://www.cdc.gov/yersinia/healthcare.html. Accessed on 15-10-2018
21. Centers for Disease Control and Prevention (CDC) (2016) Yersinia enterocolitica (Yersiniosis): diagnosis and treatment, 26 May 2016. https://www.cdc.gov/yersinia/diagnosis.html. Accessed on 15-10-2018
22. World Health Organization (2018) Campylobacter: prevention and treatment, 23 Jan 2018. http://www.who.int/news-room/fact-sheets/detail/campylobacter. Accessed on 15-10-2018
23. Centers for Disease Control and Prevention (CDC) (2017) Campylobacteriosis: diagnosis and treatment, 31 May 2017. https://wwwnc.cdc.gov/travel/yellowbook/2018/infectious-diseases-related-to-travel/campylobacteriosis. Accessed on 15-10-2018

24. Giamarellou H (2000) Therapeutic guidelines for Pseudomonas aeruginosa infections. Int J Antimicrob Agents 16(2):103–106
25. Centers for Disease Control and Prevention (CDC) (2017) Leptospirosis: prevention and treatment, 31 May 2017. https://wwwnc.cdc.gov/travel/yellowbook/2018/infectious-diseases-related-to-travel/leptospirosis. Accessed on 15-10-2018
26. Centers for Disease Control and Prevention (CDC) (2018) Leptospirosis: fact sheet for clinicians, 23 Jan 2018. https://www.cdc.gov/leptospirosis/pdf/fs-leptospirosis-clinicians-eng-508.pdf. Accessed on 15-10-2018
27. Chey WD, Wong BC, Practice Parameters Committee of the American College of Gastroenterology (2007) American College of Gastroenterology guideline on the management of Helicobacter pylori infection. Am J Gastroenterol 102(8):1808–1825
28. Centers for Disease Control and Prevention (CDC) (2017) Helicobacter pylori: diagnosis and treatment, 12 Jun 2017. https://wwwnc.cdc.gov/travel/yellowbook/2018/infectious-diseases-related-to-travel/helicobacter-pylori. Accessed on 15-10-2018
29. Centers for Disease Control and Prevention (CDC) (2017) Poliomyelitis: prevention and treatment, 31 May 2017. https://wwwnc.cdc.gov/travel/yellowbook/2018/infectious-diseases-related-to-travel/poliomyelitis. Accessed on 15-10-2018
30. World Health Organization (2017) Poliomyelitis: immunization, vaccines and biologicals, 9 Feb 2017. https://www.who.int/immunization/diseases/poliomyelitis/en/. Accessed on 15-10-2018
31. Parashar UD, Nelson EA, Kang G (2013) Diagnosis, management, and prevention of rotavirus gastroenteritis in children. BMJ 347:f7204. https://doi.org/10.1136/bmj.f7204
32. World Health Organization (2018) Rotavirus: immunization, vaccines and biologicals, Nov 2018. https://www.who.int/immunization/diseases/rotavirus/en/. Accessed on 15-10-2018
33. Centers for Disease Control and Prevention (CDC) (2018) Adenoviruses: prevention & treatment, 26 Apr 2018. https://www.cdc.gov/adenovirus/hcp/prevention-treatment.html. Accessed on 15-10-2018
34. Centers for Disease Control and Prevention (CDC) (2017) Hepatitis A: prevention & treatment, 29 Sept 2017. https://wwwnc.cdc.gov/travel/yellowbook/2018/infectious-diseases-related-to-travel/hepatitis-a. Accessed on 15-10-2018
35. World Health Organization (2018) Hepatitis A: prevention & treatment, 19 Sept 2018. http://www.who.int/news-room/fact-sheets/detail/hepatitis-a. Accessed on 15-10-2018
36. World Health Organization (2018) Hepatitis E: prevention & treatment, 19 Sept 2018. http://www.who.int/news-room/fact-sheets/detail/hepatitis-e. Accessed on 15-10-2018
37. Centers for Disease Control and Prevention (CDC) (2017) Norovirus: prevention & treatment, 31 May 2017. https://wwwnc.cdc.gov/travel/yellowbook/2018/infectious-diseases-related-to-travel/norovirus. Accessed on 15-10-2018
38. Davies AP, Chalmers RM (2009) Cryptosporidiosis. BMJ 339:b4168. https://doi.org/10.1136/bmj.b4168
39. Centers for Disease Control and Prevention (CDC) (2015) Cryptosporidium: prevention & treatment, 20 Feb 2015. https://www.cdc.gov/parasites/crypto/treatment.html. Accessed on 15-10-2018.
40. Cope JR, Ali IK (2019) Amebiasis, Travel-Related Infectious Diseases. Centers for Disease Control and Prevention (CDC). https://wwwnc.cdc.gov/travel/yellowbook/2020/travel-related-infectious-diseases/amebiasis. Accessed on 22-07-2019
41. Minetti C, Chalmers RM, Beeching NJ, Probert C, Lamden K (2016) Giardiasis. BMJ 355:i5369. https://doi.org/10.1136/bmj.i5369
42. Centers for Disease Control and Prevention (CDC) (2017) Naegleria fowleri: primary amebic meningoencephalitis (PAM) amebic encephalitis: prevention & treatment, 3 Aug 2017. www.cdc.gov/parasites/naegleria/treatment.html. Accessed on 15-10-2018
43. Pana A, Bhimji SS (2018) Amebic Meningoencephalitis. 2017 Apr 2. StatPearls [Internet]. StatPearls Publishing, Treasure Island, FL. Available from http://www.ncbi.nlm.nih.gov/books/NBK430754/PubMed

44. Centers for Disease Control and Prevention (CDC) (2018) Ascariasis: resources for health professionals, 14 Feb 2018. https://www.cdc.gov/parasites/ascariasis/health_professionals/index.html#tx. Accessed on 16-10-2018
45. World Health Organization (2018) Overview of malaria treatment, 18 Jan 2018. https://www.who.int/malaria/areas/treatment/overview/en/. Accessed on 16-10-2018
46. Centers for Disease Control and Prevention (CDC) (2013) Treatment of malaria (guidelines for clinicians), Jul 2013. https://www.cdc.gov/malaria/resources/pdf/clinicalguidance.pdf. Accessed on 16-10-2018
47. World Health Organization (2018) Lymphatic filariasis: large-scale treatment (preventive chemotherapy), 8 Oct 2018. http://www.who.int/news-room/fact-sheets/detail/lymphatic-filariasis. Accessed on 16-10-2018
48. Centers for Disease Control and Prevention (CDC) (2018) Lymphatic filariasis: prevention & treatment, 27 Sept 2018. https://www.cdc.gov/parasites/lymphaticfilariasis/treatment.html. Accessed on 16-10-2018
49. Centers for Disease Control and Prevention (CDC) (2017) Japanese encephalitis: prevention & treatment, 31 May 2017. https://wwwnc.cdc.gov/travel/yellowbook/2018/infectious-diseases-related-to-travel/japanese-encephalitis. Accessed on 16-10-2018
50. World Health Organization (2015) Japanese encephalitis: immunization, vaccines and biologicals, 9 Apr 2015. https://www.who.int/immunization/diseases/japanese_encephalitis/en/. Accessed on 16-10-2018
51. Centers for Disease Control and Prevention (CDC) (2017) Treatment of dengue fever, 31 May 2017. https://wwwnc.cdc.gov/travel/yellowbook/2018/infectious-diseases-related-to-travel/dengue. Accessed on 16-10-2018
52. World Health Organization (2018) Revised SAGE recommendation on use of dengue vaccine: immunization, vaccines and biologicals, 19 Apr 2018. https://www.who.int/immunization/diseases/dengue/revised_SAGE_recommendations_dengue_vaccines_apr2018/en/. Accessed on 16-10-2018
53. World Health Organization (2018) Zika virus: prevention & treatment, 20 July 2018. http://www.who.int/news-room/fact-sheets/detail/zika-virus. Accessed on 16-10-2018
54. Anderson JF, Rahal JJ (2002) Efficacy of interferon alpha-2b and ribavirin against West Nile virus in vitro [letter]. Emerg Infect Dis 8:107–108
55. Centers for Disease Control and Prevention (CDC) (2018) West Nile virus: prevention & treatment, 20 Jun 2018. https://www.cdc.gov/westnile/healthcareproviders/healthCareProvidersTreatmentPrevention.html. Accessed on 16-10-2018
56. Centers for Disease Control and Prevention (CDC) (2018) Yellow fever: prevention & treatment, 22 Mar 2018. https://wwwnc.cdc.gov/travel/yellowbook/2018/infectious-diseases-related-to-travel/yellow-fever. Accessed on 16-10-2018
57. World Health Organization. Yellow fever: International travel and health. https://www.who.int/ith/vaccines/yf/en/. Accessed on 16-10-2018

12 Complementary and Alternative Medicine Treatments of Water-Associated Infectious Diseases in Alliance with Conventional Medicine Treatments

Vimal K. Maurya*, Swatantra Kumar, Amrita Haikerwal, and Shailendra K. Saxena*

Abstract

The high efficacy of CAM therapies is globally acknowledged; however, their scientific validations are yet to be determined. Currently, CAM therapies are applied for treating influenza, Japanese encephalitis, inflammatory diseases, and breast cancer. These nonconventional therapies strengthen whole immune system with minimal side effects and are also cost effective. CAM is a personalized medication which alleviates the disease by treating the underlying causes coupled with it. The usefulness of CAM has been increased recently and considered as a healthcare system, product, and practice that are not a part of conventional medicine. The name CAM (complementary and alternative medicine) has been recently changed to complementary and integrative health (CIH) in Australia and other Western countries.

Keywords

CAM · Homeopathy · Ayurveda · *Rhus toxicodendron* · Hyoscyamus · *Kali phosphoricum* · Gelsemium · Belladonna

---

*Contributed equally as first author

V. K. Maurya · S. Kumar · A. Haikerwal · S. K. Saxena (✉)
Centre for Advanced Research, Faculty of Medicine, King George's Medical University, Lucknow, India
e-mail: shailen@kgmcindia.edu

S. K. Saxena (ed.), *Water-Associated Infectious Diseases*,
https://doi.org/10.1007/978-981-13-9197-2_12

12.1 Introduction

Numerous drugs are undergoing clinical trials that evaluate their efficacy against several water-associated infections but still there is an unmet gap. This lack of effective medication has inclined people to adapt towards another important medication called complementary and alternative medicine (CAM) for decimating the menace of water-associated infections. According to National Center for Complementary and Integrative Health (NCCIH) in National Institute of Health (NIH), USA, CAM comprises diversified medical practices such as homeopathy, Ayurveda, naturopathy, healthcare systems, and products which distinguish them from the conventional medicine [1]. The high efficacy of CAM therapies is globally acknowledged; however, their scientific validations are yet to be determined. Currently, CAM therapies are applied for treating influenza, Japanese encephalitis, inflammatory diseases, and breast cancer. The global alarming state of water-associated infections has motivated numerous countries like India, Brazil, Nepal, Malaysia, few nations of Africa, Europe, and the USA to accept CAM. National Health Interview Survey in 2012 reported that 17.7% of natural products are used as complementary health approaches by American adults [2]. These nonconventional therapies strengthen the whole immune system with minimal side effects and are also cost effective. CAM is a personalized medication which alleviates the disease by treating the underlying causes associated with it [3]. Moreover, the insights into the health beliefs of individuals living in regions endemic for water-associated infections are important to understand [4]. Moreover, to educate them regarding prevention and control methods and also to make them decisive in accepting the available treatments like conventional therapeutics and CAMs as the treatment are imperative. Thus, implication of CAM therapies in alliance with conventional medicine may collectively be a better opportunity to eliminate water-associated infections globally.

12.2 Homeopathy-Based Treatment of Water-Associated Diseases

Homeopathy is "a therapeutic method of using preparations of substances whose effects when administered to healthy subjects correspond to the manifestation of the disorder in the individual patient." Homeopathy is a branch of complementary and alternative medicine in which patients are treated by administration of ultradiluted form of therapeutic substances. Homeopathy was developed by Samuel Christian Friedrich Hahnemann and is now practiced throughout the world. Homeopathy is based on two principles where "like cure like" says that patients with specific sign and symptoms can be cured by homeopathic formulations which generates similar sign and symptoms in healthy individuals. The second principle is based on retaining the biological activity after repeated dilutions even beyond Avogadro's number [5]. This principle consists of three words simplex, similimum, and minimum [6] (Fig. 12.1). Treatment for water-associated infectious diseases has been mentioned in Table 12.1.

Fig. 12.1 Benefits of homeopathic medicine

12.3 Homeopathy for Water-Associated Infectious Disease

12.3.1 Typhoid

Phyto-products have been investigated for the treatment of resistant Salmonella infections where a latex protein from *Calotropis procera* has shown to be highly effective [7]. *Baptisia tinctoria* has also been investigated at the molecular level for the treatment of typhoid which produces the similar antibody generation as the natural infection with *Salmonella* [8]. The typhoid can be treated (symptomatic treatment) with various homeopathy formulations such as *Rhus toxicodendron*, Bryonia, Arnica, Arsenicum, Cinchona, carbo vegetabilis, Lachesis, muriatic acid, kali phosphoricum, Gelsemium, phosphoric acid, and Hyoscyamus [9].

Table 12.1 Homeopathic treatment for water-associated infectious diseases

Symptoms	Diseases	Homeopathic medicine
Suited to pale miserable cachectic, flabby, and broken-down constitutions, young people who look old	Filaria, diarrhea, enterocolitis	Acidum hydroflouricum
Disorientation, tremors, even loss of consciousness, muscle weakness, numbness, paralysis, extreme fatigue, fever alternating with chills, cold with violent sneezing, coughing, trembling, and strong thirst	Amoebiasis, meningoencephalitis, gastrointestinal infection	Aconite
Profuse watery discharge with sneezing and burning of the nose and lips	Diarrhea, enterocolitis	*Allium cepa*
Insect stings, burning, stinging pains, swelling of lower eyelids, absence of thirst	Filaria, Gastritis	Apis mel
Vomiting, extreme fatigue, fever alternating with chills, cold with violent sneezing, coughing, trembling, and strong thirst	Malaria	Argentum nitricum
Upper part of body hot, lower cold	Rotaviruses, filaria, gastrointestinal infection	Arnica
Thirst for a small quantity of water at frequent intervals, restlessness mentally and physically, black watery stool, and vomiting < eating and drinking after	Malaria, dengue, hepatitis E, neurological diseases	Arsenicum album
Low fever with weakness, swollen lymph glands, neck stiffness, stomach pains, mental confusion, muscle soreness, and offensive body discharges like sweat, breath, urine, stool	Meningoencephalitis	Baptisia
High fever, dry skin, red hot, sudden onset, congestive headache, red face, throbbing, violent delirium, disposition to bite, spit, strike, and tear things	Japanese encephalitis, polio	Belladonna
Profuse perspiration, especially at night, along with severe chills and fever	Malaria, leptospirosis	Boletus
Thirst increased for large quantity of water, dry mouth and tongue, constipation with no desire for stool for several days	Hand, foot, and mouth disease, meningoencephalitis, Japanese encephalitis	*Bryonia alba*
Fever with perspiration on head, vomiting less than milk after, staring of eyes, constipation	Japanese encephalitis, typhoid, diarrhea	Calc. carb
Paralytic weakness of limbs	Gastritis, Japanese encephalitis, amoebiasis	Causticum
Teething, irritability and restlessness, colic, green slimy diarrhea	Diarrhea	Chamomilla
Fever intermittent, loss of appetite, irritability, generalized weakness	Malaria, leptospirosis, cholera	China

Table 12.1 (continued)

Symptoms	Diseases	Homeopathic medicine
Malignant fevers of a hemorrhagic type which is of putrid character, cold sweat, bloody sweat	Dengue	*Crotalus horridus*
Convulsion beginning in the thumb and fingers, tonic spasm of the body, clenching thumb and fingers	Viral meningitis, polio, gastrointestinal infection	Cuprum met
Coldness of extremities, head and face hot, heat in palms and soles, and debilitation sweat	Dengue, malaria, gastritis	Ferrum metallicum
Sneezing, sore throat, symptoms of flushing, aching, trembling, weary with heavy-aching muscles, runny nose, vertigo	Meningoencephalitis, typhoid	Gelsemium
Unhealthy skin, cracked finger tips, constipation, tinnitus, earache, sinus trouble, dandruff	Gastrointestinal infection, neurological diseases	Graphites
Venous congestion, with bruised soreness of the affected parts	Filaria, bronchitis, pneumonia, hepatitis E	*Hamamelis virginiana*
Interrupted by shrieking, vacant look, stares but does not recognize those around, sensorial depression	Diarrhea, hepatitis E, cholera	Helleborus stupor
Dry eruptions on the skin, great thickening of epidermoid layer, and exfoliation of scales	Filaria, leptospirosis	Hydrocotyle asiatica
Delirium with restlessness, talks of imaginary things, makes irrelevant answers, convulsion without consciousness	Japanese encephalitis	Hyosyamus
Painful cuts and wounds, lacerated wounds involving nerve endings	Amoebiasis, meningoencephalitis, polio	Hypericum
Mental tiredness from overwork, nervous exhaustion, nervous indigestion, loss of voice, hoarseness, giddiness from exhaustion and weakness	Dengue, malaria, gastritis	Kali phos
Protrusion of tongue, evening after sleep, hot patient	Bacillus dysentery, typhoid	Lachesis
Irritability, fear of failure, period pain, premenstrual tension, hiccup with acidity	Enteric disease, leptospirosis	Lycopodium
Thirst increased with moist mouth and tongue, reproaches others	Cryptosporidiosis, dengue	Merc sol
Sneezy cold, runny nose, vertigo, premenstrual tension, housemaid's knee	Neurological diseases, hepatitis E	Nat mur
Convulsion with consciousness, chilly patient, desire to be covered	Malaria, gastritis	Nux vomica
Constricted pupils	Bacillus dysentery	Opium
High-grade fever, continuous burning in body with thirst for cold water	Hepatitis A, diarrhea, dengue	Phosphorus

(continued)

Table 12.1 (continued)

Symptoms	Diseases	Homeopathic medicine
Great lassitude, congestion of head with vertigo, face hot and flushed, prickling sensation all over, restless at night from pain in wrists and knee, rheumatic pains, profuse perspiration	Leptospirosis	*Polyporus pinicola*
Fever without thirst, dry mouth and tongue	Hand, foot, and mouth disease	Pulsatilla
Pains relieved by motion, restlessness, intense trembling, and dryness of tongue	Dengue, bacillus dysentery	Rhus tox
Injuries to bone fractures, dislocations, sprains of wrists and ankles, eye strain, urticaria	Leptospirosis, cholera	Ruta grav
Physical and mental debility due to exertion, boils, abscesses, acne, bunions, chronic headaches, sinus trouble	Filaria	Silicea
Herpetic skin with burning sensation	Neurological diseases	Staphysagria
Delirium, biting, hitting, picking of clothes	Japanese encephalitis	Stramonium
Skin eruptions, hot patient, heat of palms and soles, sensation of burning on head	Hand, foot, and mouth disease, viral meningitis	Sulfur
Hemorrhages of black blood, profuse sweat with exhaustion, fetid stools, and diminished secretion of urine	Gastritis	Sulfuric acid

12.3.2 Bacillus Dysentery

Homeopathy has been shown to be effective in case of bacillus dysentery caused by *Shigella* spp. where Mercurius Corrosivus (merc-c) has been recommended. Apart from this symptomatic treatment for bacillus dysentery can be treated with various homeopathy formulations such as Arsenicum, cantharis, aconite [10], and sulfur.

12.3.3 Cholera

Hahnemann used arsenic to treat cholera patients but eventually not all the cholera patients responded positively to arsenic [11]. Furthermore, directions from the homeopathic treatment of cholera are based on the use of four remedies. First is strong tincture of camphor. Second is *Veratrum album* used with first decimal dilution; third is Mercurius Corrosivus, with first centesimal solution; and fourth is mother tincture of ipecacuanha [12, 13]. Apart from this symptomatic treatment for cholera can be treated with various homeopathy formulations such as Camphora [14], cuprum, Arsenicum, carbo vegetabilis, Secale, podophyllum, and calcarea carbonica.

12.3.4 Enteropathogenic *E. coli*

Coli 30K is a nosode prepared from various strains of *E. coli* bacteria, which has been shown to be effective in neonatal piglets [15]. Apart from the nosode other homeopathic formulations for *E. coli* infections are carbo vegetabilis (Carb-v.), arsenicum album (Ars.), eriodictyon gluti (Erio.), Puls, Gelsemium (Gels.), ipecacuanha (Ip.), phosphorus (Phos.), antimonium tartaricum (Ant-t.), ferrum phosphoricum (Ferr-p.), and natrum muriaticum (Nat-m.) [16].

12.3.5 *Yersinia enterocolitica*

Homeopathic preparations of arsenicum album, Cinchona, aconites, *Croton tiglium*, aloe, ipeca, and podophyllum are currently used for the treatment of gastroenteritis caused by *Yersinia enterocolitica* [17].

12.3.6 *Pseudomonas aeruginosa*

Homeopathic formulations such as *Usnea barbata* might be effective which has been shown in a study where the extract of *Usnea barbata* has been used for evaluating the antimicrobial property [18].

12.3.7 Aeromonas

Rosa damascena has been shown to be effective against Aeromonas [19].

12.3.8 Legionellosis

Symptomatic treatment for pneumonia associated with legionella infection can be treated with various homeopathy formulations such as ferrum phosphoricum, kali muriaticum, phosphorus, sanguinaria, chelidonium, antimonium tartaricum, kali carbonicum, sulfur, and iodine [20].

12.3.9 Leptospirosis

NosoLEP is a nosode developed for preventing leptospirosis. NosoLEP comprises four highly diluted strains of inactivated leptospiras: *L. interrogans serovar Canicola*, *L. interrogans serovar Copenhageni*, *L. kirschneri serovar Mozdok*, and *L. borgpetersenii serovar Ballum*. The strains were selected on the basis of the frequency of isolation (circulation rate), viability, and virulence [21].

12.3.10 *Helicobacter pylori*

In case of *Helicobacter pylori* infection, nux vomica 10c and *Calendula officinalis* 10c in a 43% ethanolic solution led to a significant reduction of *H. pylori* [22]. Apart from this symptomatic homeopathic treatment of gastritis caused by *H. pylori* can be done with various homeopathy formulations such as lycopodium, phosphorus, arsenic album, *Veratrum album*, *Bryonia alba*, belladonna, graphites, carbo vegetabilis, argentum nitricum, mercurius, kali bichromicum, China officinalis, sulfur, kali carbonicum, pulsatilla, and nux vomica [23].

12.3.11 Polio Virus

The preventive homeopathic medication for polio is *Lathyrus sativus* which strengthens the immune system before the exposure of the disease. Other than this, homeopathic formulations such as calcarea phos 12X, natrum muriaticum, 12X, kalium phos 12X, *Emblica officinalis*, Gelsemium 6C, and Hypericum 12X are also administered to patients with poliomyelitis [24]. In the early stages of polio, the best medication is belladonna, Physostigma, and Eupatorium [25].

12.3.12 Coxsackie Virus A

Coxsackie virus A is the causative agent for hand, foot, and mouth disease characterized by small red spots developing into blisters on palm, soles, and oral cavity [26]. Pulsatilla is given as a prophylactic homeopathic medication to prevent the disease.

12.3.13 Coxsackie Virus B

The homeopathic treatments available for pleurodynia are *Ranunculus bulbosus*, Colchicum, aconite, Cimicifuga, Gaultheria, Arnica, Rhus R, Bryonia, and Guaiacum [27].

12.3.14 Echoviruses

The homeopathic medications for meningitis are aconite, apis mellifica, belladonna, Bryonia, Camphora, Cina maritime, Cicuta, Gelsemium, cuprum, *Helleborus niger*, *Podophyllum peltatum*, sulfur, tuberculinum, calcarea carbonica, bacillinum, *Veratrum viride*, zincum metallicum, and *Baptisia tinctoria* [28].

12.3.15 Rotaviruses

Homeopathic medications derived from plant sources are arsenicum album, *Veratrum album*, podophyllum [29], aconite, aloe socotrina, *Cinchona officinalis*, *Croton tiglium*, ipeca, mercurius corrosivus, mercurius solubilis, chamomilla, and calcarea carbonica [30].

12.3.16 Adenoviruses

The homeopathic treatment for adenoviruses includes *Vincetoxicum hirundinaria*, sulfur (Sulf.), and Engystol [15]. Other studies are suggesting the effectiveness of *Aconitum napellus*, *Bryonia alba*, *Eupatorium perfoliatum*, *Lachesis muta*, and phosphorus which are commonly recommended as homeopathic complex Gripp-Heel [31].

12.3.17 Hepatitis A Virus

Hepatitis A virus causes an acute infectious disease, hepatitis A, which is characterized by vomiting, dark urine, and jaundice. The homeopathic treatment for hepatitis A involves Chelidonium, carduus marianus, Podophyllum, Bryonia [32], and lycopodium [33].

12.3.18 Hepatitis E Virus

The homeopathic treatment for hepatitis E includes Chelidonium, carduus marianus (milk thistle), Thuja, *Lycopodium clavatum*, and belladonna [34].

12.3.19 Norovirus

Norovirus is contagious and also known as winter vomiting bug, characterized by diarrhea and inflammation in inner lining of stomach [35]. Diarrhea caused by Norovirus can be treated by arsenicum album, calcarea carbonica, chamomilla, Podophyllum, and sulfur [36].

12.3.20 *Cryptosporidium hominis*

Arsenic album and camphor can be used for the treatment of diarrhea caused by *Cryptosporidium hominis* [15].

12.3.21 *Entamoeba histolytica*

The homeopathic medications for amoebiasis are *Pulsatilla pratensis*, mercurius solubilis, mercurius corrosivus, aloe socotrina, calcarea carbonica, nux vom., and baryta carbonica [35].

12.3.22 *Giardia lamblia*

Giardia lamblia is the causative agent of a parasitic infection giardiasis, characterized by anorexia, weakness, diarrhea, bloating, abdominal cramps, and weight loss [37].

12.3.23 *Naegleria fowleri*

The homeopathic medications for primary brain meningoencephalitis are aconite, apis mellifica, *Baptisia tinctoria*, belladonna, Bryonia, Camphora, Cina maritime, Cicuta, and Gelsemium.

12.3.24 *Ascaris lumbricoides*

The homeopathic medications for *Ascaris lumbricoides* are sepia, carbo veg, mercurius, baryta carbonica, Hyoscyamus, Silicea, nux vomica, platina, *Chelone glabra*, *Urtica urens*, *Valeriana officinalis*, and *Allium sativum* [38].

12.3.25 Malaria

In homeopathy drugs are used according to the sign and symptoms produced during disease conditions. So in case of malaria all the available drugs are given according to the symptoms and severity of disease. The most effective homeopathic medicines available for the treatment of malaria are *Cinchona officinalis* (Chin.) and *Chelidonium majus* (Chel.) [39]. The most common prescribed homeopathic medicines for malaria are arsenicum album, China, *Eupatorium perfoliatum*, natrum muriaticum, nux vomica, pulsatilla, Rhus toxicodendron, and sulfur [40].

Among all the drugs, China is the most prescribed and efficient homeopathic medicine for malaria at any stage of progression. In a mice model, effects of homeopathic medications *Eupatorium perfoliatum* and arsenicum album on parasitemia of *Plasmodium berghei*-infected persons have been elucidated [41]. A combination of arsenicum album and China is found to be the most successful medicine for malaria with high fever (heat stage), chill in the morning with debilitating night sweats, malaria with high temperature that returns every week, and high restlessness associated with heat stage that gets worse after midnight.

12.3.26 Filaria

Homeopathic medicines which are commonly used for the treatment of filaria are acidum hydroflouricum (suited to pale miserable cachectic, flabby, and broken-down constitutions, young people who look old); *Hamamelis virginiana* (venous congestion, with bruised soreness of the affected parts); arsenicum album: prevents infections such as filariasis; cedron (found particularly useful in tropical, damp, warm marshy countries, found very effective in filarial infection); and hydrocotyle asiatica (prescribed for dry eruptions on the skin, thickening of epidermoid layer, and exfoliation of scales). In a single-blind 69-month follow-up study the role of Rhus tox, similimum, sulfur, Bronia, and Thuja has been shown to be effective in preventing filariasis [42]. Similarly, in another study the role of China in combination with filarinum (nosode) has been shown to be effective in the treatment of lymphatic filariasis [43].

12.3.27 Japanese Encephalitis

Various homeopathic formulations especially belladonna have been tested for the treatment of Japanese encephalitis virus infection [44]. Homeopathy treatment of JE is based on the symptoms where a range of homeopathic formulations have been suggested for specific symptoms like sensorial depression, convulsion without consciousness, sensation of burning on head, fever intermittent, loss of appetite, irritability, generalized weakness, talks of imaginary things, making irrelevant answers, convulsion beginning in the thumb and fingers, and tonic spasm of the body [45]. The formulations are as follows: belladonna, stramonium, cuprum metallicun, sulfur, *Bryonia alba*, arsenicum album, *Cinchona officinalis*, *Helleborus niger*, nux vomica, causticum, and mercurius solubilis [46].

12.3.28 Dengue

Homeopathic treatment for dengue is individualized and selection of homeopathic remedy mainly depends upon the individual response to infection, severity of disease, and clinical presentation of case and acts as a preventive measure as well as a curative for patients already suffering from dengue infection [47]. Homeopathic medicines have significant potential to reduce the intensity of fever, headache, body ache, weakness, loss of appetite, nausea, or one-time symptoms associated with dengue fever. The most functional homeopathic medicines that have been found very helpful in treating dengue fever cases are given here [48]. *Carica papaya* is recommended for increasing platelet count made from papaya fruit and seeds. *Eupatorium perfoliatum* is prescribed for dengue fever with bone-breaking pains. Conditions like nausea, vomiting of bile at the end of chill or heat stage, great soreness, and aching in bones are also treated with *Eupatorium perfoliatum* during dengue virus infection. Phosphorus: prescribed for hemorrhagic dengue fever, burning, and numbness of the hands, arms, and feet [49]. Rhus tox can be used for mainly pains relieved by motion, restlessness, intense trembling, and dryness of tongue. Bryonia can be used for dengue fever with intense thirst, sour sweat after slight exertion, and dry cough.

The following homeopathic formulations can be used to treat dengue: arsenic, belladonna, *Eupatorium perf*, Gelsemium, *Crotalus horridus*, ferrum met, Hamamelis, and Secale Cor are few of the many that help control bleeding and raise the platelet counts and also help to treat dengue infection in patients [50].

12.3.29 Zika

Homeopathic medicines like *Eupatorium perfoliatum*, Staphysagria, belladonna, *Ledum palustre*, Rhus tox, and *Polyporus pinicola* can be effectively used in Zika virus infection treatment [51].

12.3.30 West Nile

In homeopathy there is no as such treatment available for West Nile. However, further studies are required to ascertain the role of other homeopathic medicines for West Nile.

12.3.31 Yellow Fever

The following homeopathy medicines are effective for the treatment of several symptoms of yellow fever such as arsenic album, aconitc, cadmium Sulf, cantharis, carbo veg, belladonna, Crotalus H, *Allium cepa*, nux vomica, Antim Tart, argentum nitricum, Bryonia, calcarea carb, camphor, capsicum, and sulfuric acid. The following homeopathy medicine is effective in several symptoms of yellow fever: *Crotalus horridus*. However, further studies are required to ascertain the role of other homeopathic medicines for yellow fever.

12.4 Ayurveda Treatment for Water-Associated Infectious Disease

12.4.1 Typhoid

Anti-salmonella activity of herbal antityphoid preparation of various drugs like *Nauclea latifolia, Morinda lucida, Paullinia pinnata, Vernonia amygdalina, Cassia alata, Phyllanthus fraternus, Azadirachta indica, Mangifera indica,* and *Carica papaya* has been investigated and reported to be efficacious in the treatment of typhoid fever [52].

12.4.2 Bacillus Dysentery

In Ayurveda, various formulations of astringent herbs like blackberry leaf or raspberry leaf, Carob powder, and extract of bilberry and agrimony are used for the treatment of bacillus dysentery. Similarly, quercetin, chamomile, marshmallow root, and slippery elm powder are given to reduce inflammation [53].

12.4.3 Cholera

The Ayurvedic treatment for cholera helps alleviate the discomfort during the initial stages. Various "rasa" formulations such as Agnikumara rasa, Seetabhanjana rasa, Bolabadhavati, and Vishamajwarantakaloha have been used to cure cholera. Apart from rasa formulations, various home remedies like bitter gourd, lemon, onion, and nutmeg may also be prescribed for the treatment of cholera.

12.4.4 Enteropathogenic *E. coli*

Herbs known for the management diarrhea caused by enteropathogenic *E. coli* are called anti-adhesion herbs (*Aegle marmelos* unripe fruit, pulsatilla powder, *Juniperus* spp., *Salvia officinalis, Punica granatum, Terminalia chebula, Ocimum sanctum, Cinnamomum cassia, Azadirachta indica*) [54].

12.4.5 Aeromonas

Phyllanthus emblica is an Ayurvedic drug and shown to be effective against Aeromonas infection [55]. Similarly, *Lactobacillus plantarum* which is isolated from Ayurvedic medicine Kutajarista ameliorates in vitro cellular damage caused by Aeromonas [56].

12.4.6 Legionellosis

In Ayurveda, drugs like *Momordica charantia*, *Gymnema sylvestre*, and *Swertia chirata* are used for the treatment of Legionella infection.

12.4.7 Coxsackie Viruses

In Ayurveda, various preparations of *Glycyrrhiza glabra, Asparagus racemosus, Tinospora cordifolia,* and *Cyperus rotundus* are useful for the treatment of infection caused by Coxsackie virus A. According to ancient Ayurvedic texts the formulations indicated in mainly vatapittaj like Sarivadigana, Padmakadigana, Pittanashak gana, and Nyagrodhadi gana can be used in this disease as well as kalpas indicated in Masurika and Visphota chikitsa like khadiras tak, guduchyadi kwath, and patoladi kwath and locally dashanglepa and gauradhyaghrita can be used [57].

12.4.8 *Yersinia enterocolitica*

The antimicrobial activity of aqueous extract of Arjuna, Ashwagandha, Puthkanda, and Shalampanja was tested and it is reported that all the herbal medicines are found to be effective against *Yersinia enterocolitica* [58].

12.4.9 *Campylobacter jejuni*

The ethanol plant extracts of *Zingiber officinale*, *Capsicum annuum*, and *Glycyrrhiza glabra* have been investigated for anti-adhesion effects against *C. jejuni* and may be an alternative approach for the treatment of gastroenteritis [59].

12.4.10 *Pseudomonas aeruginosa*

In Ayurveda, various preparations of honey are used to treat complications caused by *Pseudomonas aeruginosa* [60].

12.4.11 Leptospirosis

The Ayurvedic treatment of leptospirosis is aimed at treating the symptoms and preventing complications like liver and kidney failure. Medicines like Tribhuvan-Kirti, Laxmi-Narayan-Ras, Sukshma-Triphala, and Maha-Sudarshan-Churna are used to treat fever. Headache and body ache can be treated using medicines like Triphala-Guggulu, Yograj-Guggulu, Maha-Rasnadi-Guggulu, Pathyadi-Qadha, Laghu-Sutshekhar-Ras, and Godanti-Bhasma.

12.4.12 *Helicobacter pylori*

In Ayurveda, various plant drugs such as curcumin, ginger, Cinnamomum, *Curcuma amada*, and berberine are given to treat *Helicobacter pylori* infection [61].

12.4.13 Rotaviruses

In Ayurveda, various herbs used in the treatment of diarrhea caused by Rotaviruses are *Terminalia arjuna, Acacia arabica, Aegle marmelos, Ficus benghalensis, Butea monosperma, Anethum sowa*, and *Emblica officinalis.*

12.4.14 Hepatitis A

In Ayurveda, a decoction preparation known as kashaya was found to be a commonly prescribed medicine for the treatment of hepatitis and consisted of Triphala, a herbal rasayana formula prepared from three equal parts of *Terminalia chebula*, *Emblica officinalis*, and *tanikka Terminalia bellirica* [33].

12.4.15 *Cryptosporidium hominis*

In Ayurveda, bitter foods such as bitter melon are excellent deterrents for parasites and natural vermifuges such as garlic, asafoetida, turmeric, ajwain, cumin, and coriander that are used for the treatment of *Cryptosporidium hominis* infection.

12.4.16 *Entamoeba histolytica*

Various natural products like quercetin, kaempferol, rutin, myricetin, hesperetin, naringenin delphinidin, cyanidin, pelargonidin, petunidin, peonidin, and malvidin are investigated against *Entamoeba histolytica* for their antiparasitic activity [62].

12.4.17 *Giardia lamblia*

In Ayurveda, aqueous extracts of *Anethum graveolens* and *Krameria cytisoides* are used for the treatment of giardiasis [63].

References

1. National Center for Complementary and Integrative Health (NCCIH) (2018) Complementary, alternative, or integrative health: What's in a name? U.S. Department of Health & Human Services, National Institutes of Health, USA.gov, Bethesda, MD. https://nccih.nih.gov/health/integrative-health. Accessed on 12-3-2018
2. Barnes PM, Bloom B, Nahin RL (2008) Complementary and alternative medicine use among adults and children: United States, 2007. Natl Health Stat Report (12):1–23
3. Saxena SK, Chitti SV, Gadugu S (2016) Complementary and alternative medicine in alliance with conventional medicine for influenza therapeutics and prevention. Future Virol 11(10):661–664
4. Wong LP, Abubakar S (2013) Health beliefs and practices related to dengue fever: a focus group study. PLoS Negl Trop Dis 7(7):1–7
5. Loudon I (2006) A brief history of homeopathy. J R Soc Med 99(12):607–610
6. Kuzeff RM (2012) A review of use of enantiomers in homeopathy. ISRN Toxicol 2012:575292
7. Oliveira RS, Figueiredo IS, Freitas LB, Pinheiro RS, Brito GA, Alencar NM, Ramos MV, Ralph MT, Lima-Filho JV (2012) Inflammation induced by phytomodulatory proteins from the latex of Calotropis procera (Asclepiadaceae) protects against Salmonella infection in a murine model of typhoid fever. Inflamm Res 61(7):689–698

8. Banerji P, Banerji P, Das GC, Islam A, Mishra SK, Mukhopadhyay S (2012) Efficacy of Baptisia tinctoria in the treatment of typhoid: its possible role in inducing antibody formation. J Complement Integr Med 9(1):15
9. Lee KJ, Yeo MG (2016) Homeopathic Rhus toxicodendron has dual effects on the inflammatory response in the mouse preosteoblastic cell line MC3T3-e1. Homeopathy 105(1):42–47
10. Prasad SK, Jain D, Patel DK, Sahu AN, Hemalatha S (2014) Antisecretory and antimotility activity of Aconitum heterophyllum and its significance in treatment of diarrhea. Indian J Pharmacol 46(1):82–87
11. Teixeira MZ (2014) 'Paradoxical pharmacology': therapeutic strategy used by the 'homeopathic pharmacology' for more than two centuries. Int J High Dilution Res. 13(49):207–226
12. Kidd J (1866) Directions for the homeopathic treatment of cholera. Edward Gould & Son, London, pp 1–16. https://collections.nlm.nih.gov/ext/cholera/PDF/101191777.pdf. Accessed on 12-3-2018
13. Gaucher C, Jeulin D, Peycru P, Pla A, Amengual C (1993) Cholera and homeopathic medicine the Peruvian experience. Br Homoeopath J. 82:155–163
14. Thielman NM, Guerrant RL (2004) Clinical practice. Acute infectious diarrhea. N Engl J Med 350(1):38–47
15. Jacobs J, Guthrie BL, Montes GA, Jacobs LE, Mickey-Colman N, Wilson AR, Digiacomo R (2006) Homeopathic combination remedy in the treatment of acute childhood diarrhea in Honduras. J Altern Complement Med 12(8):723–732
16. Camerlink I, Ellinger L, Bakker EJ, Lantinga EA (2010) Homeopathy as replacement to antibiotics in the case of Escherichia coli diarrhoea in neonatal piglets. Homeopathy 99(1):57–62
17. Velkers FC, Te Loo AJ, Madin F, van Eck JH (2005) Isopathic and pluralist homeopathic treatment of commercial broilers with experimentally induced colibacillosis. Res Vet Sci 78(1):77–83
18. Siddique F, Iqbal A, Sabir H, Manzoor M, Hussain I, Mahmood MS, Ahmad SI (2015) Eco-epidemiology and control of waterborne gastroenteritis: a review. J Food Process Technol 6(8):1–4, 477. https://doi.org/10.4172/2157-7110.1000477
19. Madamombe IT, Afolayan AJ (2003) Evaluation of antimicrobial activity of extracts from South African Usnea barbata. Pharma Bio 41(3):199–202
20. Vasanthakumar D, Aruna D, Mujeebunisa M, Saravanan S, Divya V (2015) Antibacterial activity of Rosa damascene petal extracts against the fish pathogen Aeromonas hydrophila. Euro J Exp Bio 5(8):56–59
21. Nossaman N (2017) Five cases of pneumonia cured with homeopathy: homeopathic medicine clinical snapshots. Am J Homeo Med 110(3):36–41
22. Bracho G, Varela E, Fernández R, Ordaz B, Marzoa N, Menéndez J, García L, Gilling E, Leyva R, Rufín R, de la Torre R, Solis RL, Batista N, Borrero R, Campa C (2010) Large-scale application of highly-diluted bacteria for Leptospirosis epidemic control. Homeopathy 99(3):156–166
23. Hofbauer R, Pasching E, Moser D, Frass M (2010) Heparin-binding epidermal growth factor expression in KATO-III cells after Helicobacter pylori stimulation under the influence of strychnos Nux vomica and Calendula officinalis. Homeopathy 99(3):177–182
24. Maji AK, Banerji P (2017) Strychnos nux-vomica: a poisonous plant with various aspects of therapeutic significance. J Basic Clin Pharma 8:87–103
25. Mikawlrawng K, Rani R, Kumar S, Bhardwaj AR, Prakash G (2017) Anti-paralytic medicinal plants—review. J Tradit Complement Med 8(1):4–10
26. Visintini Jaime MF, Campos RH, Martino VS, Cavallaro LV, Muschietti LV (2013) Antipoliovirus activity of the organic extract of Eupatorium buniifolium: isolation of Euparin as an active compound. Evid Based Complement Alternat Med 2013:402364
27. Harris PNA, Wang AD, Yin M, Lee CK, Archuleta S (2014) Atypical hand, foot, and mouth disease: eczema coxsackium can also occur in adults. Lancet Infect Dis 14(11):1043
28. Ozçelik B, Kartal M, Orhan I (2011) Cytotoxicity, antiviral and antimicrobial activities of alkaloids, flavonoids, and phenolic acids. Pharm Biol 49(4):396–402

29. Galabov AS, Nikolaeva-Glomb L, Mukova L, Ajvazova E, Remichkova M, Georgieva I, Cseuz M (2017) Large-scale antiviral activity screening of a triplex mixed herbal extract. Int J Herbal Med 5(5):169–176
30. Kayne S, Rafferty A (1994) The use of Arsenicum album 30c to complement conventional treatment of neonatal diarrhoea ('scours') in calves. Br Homoeopath J 83:202–204
31. Clausen J, van Wijk R, Albrecht H (2010) Infection models in basic research on homeopathy. Homeopathy 99(4):263–270
32. Rabe A (2003) A comparison of homeopathic and conventional therapies. Der Allgemeinarzt 25(20):1522–1529
33. Chandrakumar A, Xavier A, Xavier A, Manakkadiyil A, Reghu A, Thomas L (2016) Implications of traditional medicine in the treatment of Hepatitis A in Kerala. J Tradit Complement Med 6(3):316–320
34. Conti E, De Checchi G, Mencarelli R, Pinato S, Rovere P (2008) Lycopodium similiaplex-induced acute hepatitis: a case report. Eur J Gastroenterol Hepatol 20(5):469–471
35. Ganganagar S (2017) A clinical study of amoebic dysentery and its homoeopathic management. IOSR J Pharm Biol Sci. 12(1):98–102
36. The inexorable progress of norovirus (2013) Lancet Infect Dis 13(2):97
37. Minetti C, Chalmers RM, Beeching NJ, Probert C, Lamden K (2016) Giardiasis. BMJ 355:5369
38. Campos R, Amato VN, Castanho RE, Moreira AA, Pinto PL (1990) Treatment of ascaridiasis with garlic (Allium sativum). Revista do Hospital das Clinicas 45(5):213–215
39. Rajan A, Bagai U (2013) Antimalarial potential of China 30 and Chelidonium 30 in combination therapy against lethal rodent malaria parasite: Plasmodium berghei. J Complement Integr Med 10(1):89–96
40. van Erp VM, Brands M (1996) Homoeopathic treatment of malaria in Ghana: open study and clinical trial. Br Homoeopath J. 85(2):66–70
41. Lira-Salazar G, Marines-Montiel E, Torres-Monzón J, Hernández-Hernández F, Salas-Benito JS (2006) Effects of homeopathic medications Eupatorium perfoliatum and Arsenicum album on parasitemia of Plasmodium berghei-infected mice. Homeopathy 95(4):223–228
42. Kumar A, Mishra N (1994) Effect of homoeopathic treatment on filariasis: a single blind 69-months follow-up study in an endemic village in Orissa. Br Homoeopath J. 83(4):216–219
43. Rudra SK, Sukul NC, Sukul A (2009) Antifilarial effect of two homeopathic potencies against human filariasis. Clin Exp Homeopathy. 1:2
44. Bandyopadhyay B, Das S, Sengupta M, Saha C, Das KC, Sarkar D, Nayak C (2010) Decreased intensity of Japanese encephalitis virus infection in chick chorioallantoic membrane under influence of ultradiluted Belladonna extract. Am J Infect Dis 6(2):24–28
45. Saxena SK, Singh M, Pathak AK, Mathur A (2006) Reply to 'Encephalitis outbreak finds Indian officials unprepared'. Nat Med 12(3):269–270
46. Manchanda RK, Oberai P, Roja V, Singh S, Singh N, Khan T et al (2015) Evaluation of homoeopathic medicines as add-on to institutional management protocol in acute encephalitis syndrome: an exploratory observational comparative study. Indian J Res Homoeopathy 9(1):34–41
47. Marino R (2008) Homeopathy and collective health: the case of dengue epidemics. Int J High Dilution Res 7(25):179–185
48. Saeed-ul-Hassan S, Tariq I, Khalid A, Karim S (2013) Comparative clinical study on the effectiveness of homeopathic combination remedy with standard maintenance therapy for dengue fever. Trop J Pharm Res 12(5):767–770
49. dos Santos AL, Perazzo FF, Cardoso LG, Carvalho JC (2007) In vivo study of the anti-inflammatory effect of Rhus toxicodendron. Homeopathy 96(2):95–101
50. Jacobs J, Fernandez EA, Merizalde B, Avila-Montes GA, Crothers D (2007) The use of homeopathic combination remedy for dengue fever symptoms: a pilot RCT in Honduras. Homeopathy 96(1):22–26

51. Saxena SK, Elahi A, Gadugu S, Prasad AK (2016) Zika virus outbreak: an overview of the experimental therapeutics and treatment. Virus 27(2):111–115
52. Koffuor GA, Abruquah AA, Audu R, Amoah J, Agwah D (2016) Patronage and perceived efficacy of herbal antityphoid preparations, and anti-Salmonella activity of a herbal preparation used in Ghana. J Appl Pharm Sci 6(3):1–7
53. Zhang HY, Lu SF, Xiao N (2009) Effect of warming moxibustion on Shenque acupoint for the treatment of acute diarrhea in children with infantile cerebral palsy. Chin J Integr Med 15(6):454–457
54. Brijesh S, Daswani P, Tetali P, Antia N, Birdi T (2009) Studies on the antidiarrhoeal activity of Aegle marmelos unripe fruit: validating its traditional usage. BMC Complement Altern Med 9:47
55. Anto VJ, Balasubramanian V (2015) Therapeutic effect of Phyllanthus emblica on disease induced common carp Cyprinus carpio by Aeromonas hydrophila. Int J Zool Res 11:96–101
56. Kumar H, Rangrez AY, Dayananda KM, Atre AN, Patole MS, Shouche YS (2011) Lactobacillus plantarum (VR1) isolated from an Ayurvedic medicine (Kutajarista) ameliorates in vitro cellular damage caused by Aeromonas veronii. BMC Microbiol 11:152
57. Chen X, Wang C, Xu L, Chen X, Wang W, Yang G, Tan RX, Li E, Jin Y (2013) A laboratory evaluation of medicinal herbs used in China for the treatment of hand, foot, and mouth disease. Evid Based Complement Alternat Med 2013:504563
58. Sharma S, Kaushik R, Sharma P, Sharma R, Thapa A, Kp I (2016) Antimicrobial activity of herbs against Yersinia enterocolitica and mixed microflora. Ann Univ Dunarea de Jos of Galati Fascicle VI—Food Technol 40(2):119–134
59. Bensch K, Tiralongo J, Schmidt K, Matthias A, Bone KM, Lehmann R, Tiralongo E (2011) Investigations into the antiadhesive activity of herbal extracts against campylobacter jejuni. Phytother Res 25(8):1125–1132
60. Shenoy VP, Ballal M, Shivananda PG, Bairy I (2012) Honey as an antimicrobial agent against Pseudomonas aeruginosa isolated from infected wounds. J Global Infect Dis 4(2):102
61. Mishra M, Mishra S (2014) Role of Ayurveda in the management of Helicobacter pylori infection. American Journal of Pharmacy & Health Research. 2: 2321–3647
62. Martínez-Castillo M, Pacheco-Yepez J, Flores-Huerta N, Guzmán-Téllez P, Jarillo-Luna RA, Cárdenas-Jaramillo LM, Campos-Rodríguez R, Shibayama M (2018) Flavonoids as a natural treatment against Entamoeba histolytica. Front Cell Infect Microbiol 8:209
63. Sahib AS, Mohammed IH, Sloo SA (2014) Antigiardial effect of Anethum graveolens aqueous extract in children. J Intercult Ethnopharmacol. 3(3):109–112

MIX
Papier aus verantwortungsvollen Quellen
Paper from responsible sources
FSC® C105338

If you have any concerns about our products,
you can contact us on
ProductSafety@springernature.com

In case Publisher is established outside the EU,
the EU authorized representative is:
Springer Nature Customer Service Center GmbH
Europaplatz 3, 69115 Heidelberg, Germany

Printed by Libri Plureos GmbH
in Hamburg, Germany